SAFETY SYMBOLS

SAFETY SYMBOLS	HAZARD	EXAMPLES	PRECAUTION	REMEDY
DISPOSAL	Special disposal procedures need to be followed.	certain chemicals, living organisms	Do not dispose of these materials in the sink or trash can.	Dispose of wastes as directed by your teacher.
BIOLOGICAL	Organisms or other biological materials that might be harmful to humans	bacteria, fungi, blood, unpreserved tissues, plant materials	Avoid skin contact with these materials. Wear mask or gloves.	Notify your teacher if you suspect contact with material. Wash hands thoroughly.
EXTREME TEMPERATURE	Objects that can burn skin by being too cold or too hot	boiling liquids, hot plates, dry ice, liquid nitrogen	Use proper protection when handling.	Go to your teacher for first aid.
SHARP OBJECT	Use of tools or glassware that can easily puncture or slice skin	razor blades, pins, scalpels, pointed tools, dissecting probes, broken glass	Practice common-sense behavior and follow guidelines for use of the tool.	Go to your teacher for first aid.
FUME	Possible danger to respiratory tract from fumes	ammonia, acetone, nail polish remover, heated sulfur, moth balls	Make sure there is good ventilation. Never smell fumes directly. Wear a mask.	Leave foul area and notify your teacher immediately.
ELECTRICAL	Possible danger from electrical shock or burn	improper grounding, liquid spills, short circuits, exposed wires	Double-check setup with teacher. Check condition of wires and apparatus.	Do not attempt to fix electrical problems. Notify your teacher immediately.
IRRITANT	Substances that can irritate the skin or mucus membranes of the respiratory tract	pollen, moth balls, steel wool, fiberglass, potassium permanganate	Wear dust mask and gloves. Practice extra care when handling these materials.	Go to your teacher for first aid.
CHEMICAL	Chemicals that can react with and destroy tissue and other materials	bleaches such as hydrogen peroxide; acids such as sulfuric acid, hydrochloric acid; bases such as ammonia, sodium hydroxide	Wear goggles, gloves, and an apron.	Immediately flush the affected area with water and notify your teacher.
TOXIC	Substance may be poisonous if touched, inhaled, or swallowed	mercury, many metal compounds, iodine, poinsettia plant parts	Follow your teacher's instructions.	Always wash hands thoroughly after use. Go to your teacher for first aid.
OPEN FLAME	Open flame may ignite flammable chemicals, loose clothing, or hair	alcohol, kerosene, potassium permanganate, hair, clothing	Tie back hair. Avoid wearing loose clothing. Avoid open flames when using flammable chemicals. Be aware of locations of fire safety equipment.	Notify your teacher immediately. Use fire safety equipment if applicable.

Eye Safety
Proper eye protection should be worn at all times by anyone performing or observing science activities.

Clothing Protection
This symbol appears when substances could stain or burn clothing.

Animal Safety
This symbol appears when safety of animals and students must be ensured.

Radioactivity
This symbol appears when radioactive materials are used.

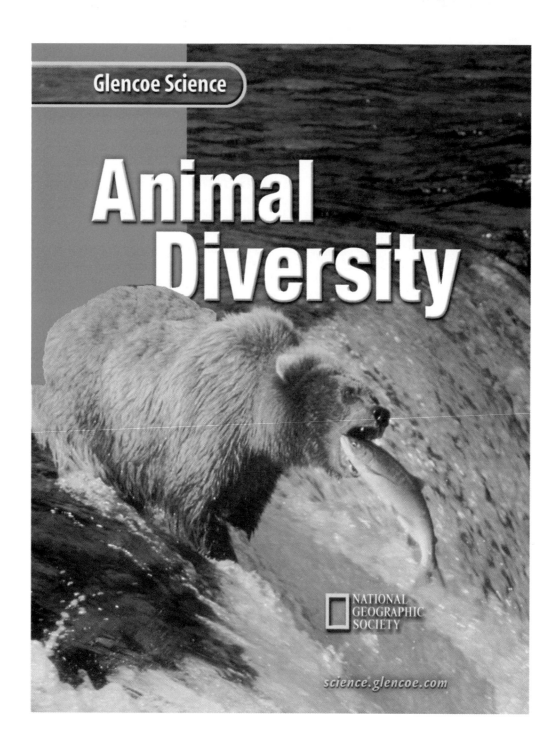

Animal Diversity

Glencoe Science

science.glencoe.com

Glencoe McGraw-Hill

New York, New York   Columbus, Ohio   Woodland Hills, California   Peoria, Illinois

Glencoe Science
Animal Diversity

Student Edition
Teacher Wraparound Edition
Interactive Teacher Edition CD-ROM
Interactive Lesson Planner CD-ROM
Lesson Plans
Content Outline for Teaching
Dinah Zike's Teaching Science with Foldables
Directed Reading for Content Mastery
Foldables: Reading and Study Skills
Assessment
    Chapter Review
    Chapter Tests
    ExamView Pro Test Bank Software
    Assessment Transparencies
    Performance Assessment in the Science Classroom
    The Princeton Review Standardized Test Practice Booklet
Directed Reading for Content Mastery in Spanish
Spanish Resources
English/Spanish Guided Reading Audio Program
Reinforcement

Enrichment
Activity Worksheets
Section Focus Transparencies
Teaching Transparencies
Laboratory Activities
Science Inquiry Labs
Critical Thinking/Problem Solving
Reading and Writing Skill Activities
Mathematics Skill Activities
Cultural Diversity
Laboratory Management and Safety in the Science Classroom
Mindjogger Videoquizzes and Teacher Guide
Interactive Explorations and Quizzes CD-ROM with
    Presentation Builder
Vocabulary Puzzlemaker Software
Cooperative Learning in the Science Classroom
Environmental Issues in the Science Classroom
Home and Community Involvement
Using the Internet in the Science Classroom

"Study Tip," "Test-Taking Tip," and the "Test Practice" features in this book were written by The Princeton Review, the nation's leader in test preparation. Through its association with McGraw-Hill, The Princeton Review offers the best way to help students excel on standardized assessments.

The Princeton Review is not affiliated with Princeton University or Educational Testing Service.

Glencoe/McGraw-Hill
A Division of The McGraw·Hill Companies

Copyright ©2002 by the McGraw-Hill Companies, Inc. All rights reserved. Except as permission under the United States Copyright Act, no part of this publication may be reproduced or distributed in any form or by any means, or stored in a database or retrieval system, without the prior written permission of the publisher.

The "Visualizing" features found in each chapter of this textbook were designed and developed by the National Geographic Society's Education Division, copyright ©2002 National Geographic Society. The name "National Geographic Society" and the yellow border rectangle are trademarks of the Society, and their use, without prior written permission, is strictly prohibited. All rights reserved.

The "Science and Society" and the "Science and History" features that appear in this book were designed and developed by TIME for Kids, a division of TIME Magazine.

Cover Images: An Alaskan brown bear catches a migrating salmon.

Send all inquiries to:
Glencoe/McGraw-Hill
8787 Orion Place
Columbus, OH 43240

ISBN 0-07-825567-8
Printed in the United States of America.
1 2 3 4 5 6 7 8 9 10 027/043 06 05 04 03 02 01

Authors

Lucy Daniel, PhD
Teacher/Consultant
Rutherfordton-Spindale High School
Rutherfordton, North Carolina

Dinah Zike
Educational Consultant
Dinah-Might Activities, Inc.
San Antonio, Texas

Consultants

Content

Jerry Jackson, PhD
Program Director Center for Science, Mathematics, and Technology Education
Florida Gulf Coast University
Fort Meyers, Florida

Dominic Salinas, PhD
Middle School Science Supervisor
Caddo Parish Schools
Shreveport, Louisiana

Safety

Sandra West, PhD
Associate Professor of Biology
Southwest Texas State University
San Marcos, Texas

Reading

Carol A. Senf, PhD
Associate Professor of English
Georgia Institute of Technology
Atlanta, Georgia

Math

Teri Willard, EdD
Department of Mathematics
Montana State University
Belgrade, Montana

Reviewers

Maureen Barrett
Thomas E. Harrington Middle School
Mt. Laurel, New Jersey

Janice Bowman
Coke R. Stevenson Middle School
San Antonio, Texas

Cory Fish
Burkholder Middle School
Henderson, Nevada

Diane Lutz
Denmark Middle School
Denmark, Wisconsin

Amy Morgan
Berry Middle School
Hoover, Alabama

Michelle Punch
Northwood Middle School
Houston, Texas

Darcy Vetro-Ravndal
Middleton Middle School of Technology
Tampa, Florida

Billye Robbins
Lomax Junior High School
LaPorte, Texas

Delores Stout
Bellefonte Middle School
Bellefonte, Pennsylvania

Series Activity Testers

José Luis Alvarez, PhD
Math/Science Mentor Teacher
Yseleta ISD
El Paso, Texas

Nerma Coats Henderson
Teacher
Pickerington Jr. High School
Pickerington, Ohio

Mary Helen Mariscal-Cholka
Science Teacher
William D. Slider Middle School
El Paso, Texas

José Alberto Marquez
TEKS for Leaders Trainer
Yseleta ISD
El Paso, Texas

Science Kit and Boreal Laboratories
Tonawanda, New York

Contents

Nature of Science: Monarch Migration—2

Introduction to Animals—6

CHAPTER 1

- **SECTION 1** Is it an animal? ... 8
- **SECTION 2** Sponges and Cnidarians ... 14
  - Activity Observing a Cnidarian ... 21
- **SECTION 3** Flatworms and Roundworms ... 22
  - NATIONAL GEOGRAPHIC Visualizing Parasitic Worms ... 26
  - Activity: Design Your Own Experiment
    Comparing Free-Living and Parasitic Worms ... 28
  - TIME *Science and History*
    Sponges ... 30

Mollusks, Worms, Arthropods, and Echinoderms—36

CHAPTER 2

- **SECTION 1** Mollusks ... 38
- **SECTION 2** Segmented Worms ... 43
- **SECTION 3** Arthropods ... 48
  - NATIONAL GEOGRAPHIC Visualizing Arthropod Diversity ... 54
  - Activity Observing a Crayfish ... 57
- **SECTION 4** Echinoderms ... 58
  - Activity What do worms eat? ... 62
  - Science and Language Arts
    "Creatures of My Mind" ... 64

Fish, Amphibians, and Reptiles—70

CHAPTER 3

- **SECTION 1** Chordates and Vertebrates ... 72
  - Activity Endotherms and Ectotherms ... 76
- **SECTION 2** Fish ... 77
- **SECTION 3** Amphibians ... 85
- **SECTION 4** Reptiles ... 90
  - NATIONAL GEOGRAPHIC Visualizing Extinct Reptiles ... 94
  - Activity: Design Your Own Experiment
    Water Temperature and the Respiration Rate of Fish ... 96
  - TIME *Science and Society*
    Venom ... 98

Contents

CHAPTER 4

Birds and Mammals—104

SECTION 1 Birds 106
    NATIONAL GEOGRAPHIC Visualizing Birds 112

SECTION 2 Mammals 114
    Activity Mammal Footprints 123
    Activity: Use the Internet
       Bird Counts 124
    Science Stats Eggciting Facts 126

CHAPTER 5

Animal Behavior—132

SECTION 1 Types of Behavior 134
SECTION 2 Behavioral Interactions 140
    NATIONAL GEOGRAPHIC Visualizing Bioluminescence 145

    Activity Observing Earthworm Behavior 149
    Activity: Model and Invent
       Animal Habitats 150
    Oops! Accidents in Science Going to the Dogs 152

Field Guides
    Insects Field Guide 162
    Feline Traits Field Guide 166

Skill Handbooks—170

Reference Handbooks
    A. Safety in the Science Classroom 195
    B. SI Metric to English, English to Metric Conversions .. 196
    C. Care and Use of a Microscope 197
    D. Diversity of Life 198

English Glossary—202

Spanish Glossary—207

Index—212

Interdisciplinary Connections/Activities

NATIONAL GEOGRAPHIC VISUALIZING

1 Parasitic Worms................26
2 Arthropod Diversity.............54
3 Extinct Reptiles................94
4 Birds.........................112
5 Bioluminescence...............145

TIME SCIENCE AND Society

3 Venom........................98

TIME SCIENCE AND HISTORY

1 Sponges.......................30

Accidents in SCIENCE

5 Going to the Dogs.............152

Science and Language Arts

2 "Creatures of My Mind"..........64

Science Stats

4 Eggciting Facts................126

Full Period Labs

1 Observing a Cnidarian...........21
   Design Your Own Experiment:
   Comparing Free-Living
   and Parasitic Worms..........28
2 Observing a Crayfish............57
   What do worms eat?............62
3 Endotherms and Ectotherms......76
   Design Your Own Experiment:
   Water Temperature and the
   Respiration Rate of Fish.......96
4 Mammal Footprints.............123
   Use the Internet: Bird Counts....124
5 Observing Earthworm Behavior..149
   Model and Invent:
   Animal Habitats..............150

EXPLORE ACTIVITY

1 Demonstrate symmetry............7
2 Examine a clam's shell..........37
3 Model how a snake hears.........71
4 Model how a bird's gizzard works.105
5 Observe how humans communicate
   without using sound..........133

Activities/Science Connections

Mini LAB

1. **Try at Home:** Modeling Animal Camouflage 10
   Observing Planarian Movements .. 24
2. Observing Metamorphosis......... 50
   Try at Home: Modeling the Strength of Tube Feet 59
3. **Try at Home:** Modeling How Fish Adjust to Different Depths.. 80
   Describing Frog Adaptations 88
4. Modeling Feather Preening 108
   Try at Home: Inferring How Blubber Insulates............. 116
5. Observing Conditioning......... 138
   Try at Home: Demonstrating Chemical Communication..... 143

Math Skills Activities

1. Using Percent to Calculate Number of Species 25
2. Calculating Percent of Elasticity 2
3. Calculating Density.............. 82

Problem Solving Activities

1. Does a mammal's heart rate determine how long it will live? . 117
5. How can you determine which animals hibernate?............ 147

Skill Builder Activities

Science Skill Handbook
Classifying: 122
Communicating: 13, 27, 61, 75, 89, 113
Comparing and Contrasting: 20, 47
Concept Mapping: 13, 27, 56, 75, 84, 89, 95, 113
Forming Hypotheses: 61
Interpreting Scientific Illustrations: 42
Making and Using Graphs: 56
Researching Information: 139
Testing a Hypothesis: 148

Math Skill Handbook
Solving One-Step Equations: 20, 95, 122, 148
Using Proportions: 47

Technology Skill Handbook
Using a Database: 84
Using an Electronic Spreadsheet: 139
Using a Word Processor: 42

Science INTEGRATION
Chemistry: 16, 144
Earth Science: 41, 95
Environmental Science: 109
Health: 9, 51, 80, 135

SCIENCE Online
Collect Data: 18
Research: 12, 46, 60, 87, 91, 110, 119, 137, 146

Test Practice: 35, 69, 103, 131, 157, 158–159

The Nature of Science

Scientific Methods

Monarch Migration

Figure 1
Eastern monarch butterflies lay their eggs on milkweed plants.

Although the beautiful black and orange wings of the monarch butterfly are a common sight during summer in the United States, as fall and colder temperatures arrive, the butterflies disappear. Each fall they begin a seasonal migration. Scientists have had some success in unlocking the mystery of monarch migration through the use of scientific methods. Through this example, you can see how each step of this scientific method contributes to reliable results that can lead to better-informed conclusions.

The monarch population that lives west of the Rocky Mountains flies to the coast of California. The eastern population of monarchs flies to the mountains of central Mexico. Sometimes they travel up to 145 km per day. Some eastern monarchs, such as those living in southern Canada, fly more than 3,200 km to reach their winter home.

Navigation

Just as astonishing as the distance traveled by these insect voyagers is their ability to find their way. Since no butterfly completes the entire round-trip, the butterflies cannot learn the route from others. So, how do butterflies that have never made the trip before find their way from Canada and the eastern United States to Mexico? This is the question that some entomologists—scientists who study insects—set out to answer.

One of the first hypotheses about how eastern monarchs navigate was that they use the Sun as a guide.

Figure 2
When they reach Mexico, eastern monarch butterflies gather in large groups.

Figure 3
Magnetite is a mineral with natural magnetic properties.

Researchers based this educated guess on other research, which showed that some migrating birds rely on the Sun to guide them. But, this theory alone failed to explain how the butterflies find their way on cloudy days.

Magnetism

Scientists later discovered that the bodies of eastern monarchs contain tiny grains of a naturally occurring, magnetic substance called magnetite, which was used to make the first directional compasses. From this discovery, scientists developed a theory that the butterflies use an internal magnetic compass to help them plot their route.

University scientists tested this hypothesis by performing an experiment. They caught some eastern monarchs during the fall migration. They divided the monarchs into three groups and exposed each group to different magnetic fields. The group exposed to Earth's normal magnetic field flew to the southwest, which is the correct direction for eastern monarchs to migrate. Those exposed to the opposite of Earth's normal magnetic field flew to the northeast. Finally, those exposed to no magnetic field fluttered about randomly.

Final Conclusions

After analyzing the results, the researchers concluded that eastern monarchs use an internal magnetic compass to navigate from Canada and the eastern United States to Mexico and back again. However, most researchers also believe the butterflies also use the Sun and landmarks, such as mountains and rivers, to make their incredible journey.

Figure 4
A magnet has oppositely charged poles.

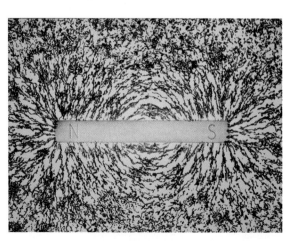

The Nature of Science

Science

Scientists learned about the migration of eastern monarch butterflies by starting with questions, such as "Where do monarchs go each fall? How do they find their way there?" Scientists use experiments and careful observations to answer questions about how the world works. When you test an idea, you are doing science.

Life science is the study of living things. In this book, you will learn about the diversity of animals and their adaptations and behaviors, such as migration, that enable them to survive.

Scientific Methods

Researchers used scientific methods to learn about how eastern monarchs navigate. Scientific methods are a series of procedures used to investigate a question scientifically.

Identifying a Question

Sometimes, scientific methods begin with identifying a question, such as "Where do eastern monarch butterflies go every autumn?" After one question has been answered, others often arise. When researchers discovered eastern monarchs migrate to Mexico, the next question was "How do the butterflies find their way?"

Forming a Hypothesis

Once a question has been identified, scientists collect information on the subject and develop a hypothesis, or educated guess.

They might read the information available on how birds migrate and use it as a basis for the hypothesis that eastern monarchs use the Sun to navigate. This idea, however, failed to explain how the butterflies find their way on cloudy days. As a result, scientists developed another hypothesis—eastern monarchs use an internal magnetic compass that enables them to maintain a course in a specific direction.

Figure 5
Sometimes, a scientist must collect data outside of the laboratory.

Testing the Hypothesis

Scientists test hypotheses to determine if they are true or false. Such tests often involve experiments, such as one where eastern monarchs were exposed to different kinds of magnetic fields.

Analyzing Results

During experiments, scientists gather a lot of information, or data. Data about the butterfly experiment included the direction that the butterflies were flying when captured, what type of magnetic field they were exposed to in the experiment, and how they reacted to that magnetic field.

Drawing a Conclusion

After data have been collected and carefully analyzed, scientists draw conclusions. Sometimes the original hypothesis is not supported by the data and scientists must start the entire process over. In the case of the eastern monarchs, researchers observed how the butterflies reacted to the magnetic fields and concluded they use an internal magnetic compass to navigate. Just how the butterflies use Earth's magnetic field to find their way is another question for scientists to answer using scientific methods.

Figure 6
Data from observations are important in science investigations.

Figure 7
Scientists hypothesize that monarchs also navigate by landmarks.

When eastern monarch butterflies reach Mexico's mountains, the insects abruptly change direction. Scientists hypothesize that the butterflies then switch to steering by landmarks, such as mountains. Describe one way scientists could test this hypothesis.

CHAPTER 1

Introduction to Animals

Did you know that some cultures classified animals according to how useful or destructive they were to humans? Some animal groups were based on their roles in myths and legends. In this chapter, you will discover how animals are classified today and learn about the relationships that exist among different groups in the animal kingdom. You also will learn how some animals—such as sponges, jellyfish, roundworms, and flatworms—live and reproduce and how they are important to humans.

What do you think?

Science Journal Look at the picture below with a classmate. Discuss what you think this might be. Here's a hint: *Can these organisms move around or are they attached to the ocean floor?* Write your answer or best guess in your Science Journal.

Explore Activity

The words *left* and *right* have meaning to you because your body has a left and a right side. But what is left or right to a jellyfish or sea star? How an animal's body parts are arranged is called symmetry. In the following activity, you will compare three types of symmetry found in animals.

Demonstrate symmetry

1. On a piece of paper, draw three shapes—a circle, a triangle with two equal sides, and a free-form shape—then cut them out.

2. Fold each shape through the center as many different ways as you can to make similar halves with each fold.

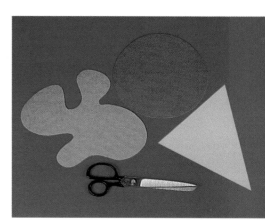

Observe

Which shapes can be folded into equal halves and which shapes cannot? Can any of the shapes be folded into equal halves more than one way? Record your answers in your Science Journal.

Before You Read

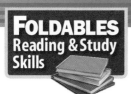

Making an Organizational Study Fold
Make the following Foldable to help you organize your thoughts about your favorite animals are how they are classified.

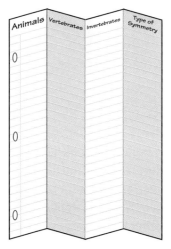

1. Place a sheet of paper in front of you so the short side is at the top. Fold both sides in then unfold.

2. Fold the paper in half from top to bottom. Then fold it in half again two more times. Unfold all the folds and trace over the vertical fold lines.

3. Label the columns *Animals, Vertebrates, Invertebrates,* and *Type of Symmetry* as shown.

4. Before you read the chapter, list your four favorite animals in the *Animal* column.

5. As you read the chapter, add information about how the four animals are classified to the table.

C ◆ 7

SECTION 1

Is it an animal?

As You Read

What You'll Learn
- **Identify** the characteristics common to most animals.
- **Determine** how animals meet their needs.
- **Distinguish** between invertebrates and vertebrates.

Vocabulary
herbivore
carnivore
omnivore
vertebrate
invertebrate
radial symmetry
bilateral symmetry

Why It's Important
Animals provide food, medicines, and companionship in your daily life.

Animal Characteristics

From microscopic worms to giant whales, the animal kingdom includes an amazing variety of living things, but all of them have certain characteristics in common. What makes the animals in **Figure 1** different from plants? Is it because animals eat other living things? Is this enough information to identify them as animals? What characteristics do animals have?

1. Animals are made of many cells. The cells are of different kinds that carry out different functions such as sensing the environment, getting rid of wastes, and reproducing.

2. Animal cells have a nucleus and specialized structures inside the cells called organelles.

3. Animals depend on other living things in the environment for food. Some eat plants, some eat other animals, and some eat plants and animals.

4. Animals digest their food. The proteins, carbohydrates, and fats in foods are broken down into simpler molecules that can move into the animal's cells.

5. Many animals move from place to place. They can escape from their enemies and find food, mates, and places to live. Animals that move slowly or not at all have adaptations that make it possible for them to take care of these needs in other ways.

6. All animals are capable of reproducing sexually. Some animals also can reproduce asexually.

Figure 1
These organisms look like plants, but they're one of the many plantlike animals that can be found growing on shipwrecks and other underwater surfaces.

8 ◆ C

Figure 2
Animals eat a variety of foods.

A Chitons eat algae from rocks.

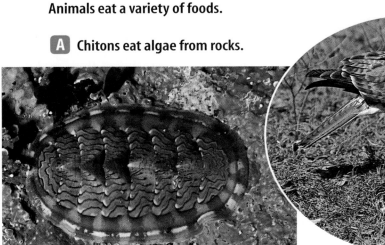

B A red-tailed hawk uses its sharp beak to tear the flesh.

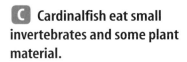

C Cardinalfish eat small invertebrates and some plant material.

How Animals Meet Their Needs

Any structure, process, or behavior that helps an organism survive in its environment is an adaptation. Adaptations are inherited from previous generations. In a changing environment, adaptations determine which individuals are more likely to survive and reproduce.

Adaptations for Obtaining Energy One of the most basic needs of animals is the need for food. All animals have adaptations that allow them to obtain, eat, and digest different foods. The chiton shown in **Figure 2A,** deer, some fish, and many insects are examples of herbivores. **Herbivores** eat only plants or parts of plants. In general, herbivores eat more often and in greater amounts than other animals because plants don't supply as much energy as other types of food.

Reading Check *Why are butterflies considered to be herbivores?*

Animals that eat only other animals, like the red-tailed hawk in **Figure 2B,** are **carnivores.** Most carnivores capture and kill other animals for food. But some carnivores, called scavengers, eat only the remains of other animals. Animal flesh supplies more energy than plants do, so carnivores don't need to eat as much or as often as herbivores.

Animals that eat plants and animals or animal flesh are called **omnivores.** Bears, raccoons, robins, humans, and the cardinalfish in **Figure 2C** are examples of omnivores.

Many beetles and other animals such as millipedes feed on tiny bits of decaying matter called detritus (dih TRI tus). They are called detritivores (dih TRI tih vorz).

Health INTEGRATION

Many animals, including humans, have microorganisms in their digestive tracts. These microorganisms are helpful in preventing harmful bacteria from growing in the intestines. In your Science Journal, infer why microorganisms are important for good health in animals.

Figure 3 The pill bug's outer covering protects it and reduces moisture loss from its body.

Physical Adaptations Some prey species have physical features that enable them to avoid predators. Outer coverings protect some animals. Pill bugs, as seen in **Figure 3,** have protective plates. Porcupines have sharp quills that prevent most predators from eating them. Turtles and many animals that live in water have hard shells that protect them from predators.

Size is also a type of defense. Large animals are usually safer than small animals. Few predators will attack animals such as moose or bison simply because they are so large.

Mimicry is an adaptation in which one animal closely resembles another animal in appearance or behavior. If predators cannot distinguish between the two, they usually will not eat either animal. The venomous coral snake and the nonvenomous scarlet king snake, shown in **Figure 4,** look alike. In some cases, this is a disadvantage for scarlet king snakes because people mistake them for coral snakes and kill them.

Reading Check *How might mimicry be an advantage and a disadvantage for an animal?*

Many animals, like the flounder in **Figure 5A,** blend into their surrounding environment, enabling them to hide from their predators. English peppered moths are brown and speckled like the lichens (LI kunz) on trees, making it difficult for their predators to see them. Many freshwater fish, like the trout in **Figure 5B,** have light bellies and dark, speckled backs that blend in with the gravelly bottoms of their habitats when they are viewed from above. Any marking or coloring that helps an animal hide from other animals is called camouflage. Some animals, like the chameleons in **Figure 5C,** have the ability to change their color depending on their surroundings.

TRY AT HOME Mini LAB

Modeling Animal Camouflage

Procedure
1. Pretend that a room in your home is the world of some fictitious animal. From **materials you can find around your home,** build a **fictitious animal** that would be camouflaged in this world.
2. Put your animal into its world and ask someone to find it.

Analysis
1. In how many places was your animal camouflaged?
2. What changes would increase its chances of surviving in its world?

Figure 4 Mimicry helps some animals survive. Compare the colors and patterns of **A** the coral snake and **B** the scarlet king snake. *What is the difference between the two snakes?*

Figure 5 Many types of animals blend with their surroundings.

A Bottom fish like this flounder, blend with the ocean floor. *Can you see the flounder in this photo?*

B A trout blends with the bottom of a stream.

C Chameleons can be especially difficult to find because they can change color to blend with their surroundings.

Predator Adaptations Camouflage is an adaptation for many predators so they can sneak up on their prey. Tigers have stripes that hide them in tall grasses. Killer whales are black on their upper surface and white underneath. When seen from above, the whale blends into the darkness of the deep ocean. The killer whale's white underside appears to be nearly the same color as the bright sky overhead when viewed from below. Adaptations such as these enable predators to hunt prey more successfully.

Behavioral Adaptations In addition to physical adaptations, animals have behavioral adaptations that enable them to capture prey or to avoid predators. Chemicals are used by some animals to escape predators. Skunks spray attacking animals with a bad-smelling liquid. Some ants and beetles also use this method of defense. When squid and octopuses are threatened, they release a cloud of ink so they can escape, as shown in **Figure 6.**

Some animals are able to run faster than most of their predators. The Thomson's gazelle can run at speeds up to 80 km/h. A lion can run only about 36 km/h, so speed is a factor in the Thomson gazelle's survival.

Traveling in groups is a behavior that is demonstrated by predators and prey. Herring swim in groups called schools that resemble an organism too large for a predator fish to attack. On the other hand, when wolves travel in packs, they can successfully hunt large prey that one predator alone could not capture.

Figure 6 An octopus's cloud of ink confuses a predator long enough for the octopus to escape.

Figure 7
Animals can be classified into two large groups. These groups can be broken down further based on different animal characteristics.

Animal Classification

Scientists have identified and named more than 1.8 million species of animals. It is estimated that there are another 3 million to 30 million more to identify and name. Animals can be classified into two major groups, as shown in **Figure 7.** All animals have common characteristics, but those in one group have more, similar characteristics because all the members of a group probably descended from a common ancestor. When a scientist finds a new animal, how does he or she begin to classify it?

Check for a Backbone To classify an animal, a scientist first looks to see whether or not the animal has a backbone. Animals with backbones are called **vertebrates.** Their backbones are made up of a stack of structures called vertebrae that support the animal. The backbone also protects and covers the spinal cord—a bundle of nerves that is connected to the brain. The spinal cord carries messages to all other parts of the body. It also carries messages from other parts of the body to the brain. Examples of vertebrates include fish, frogs, snakes, birds, and humans.

An animal without a backbone is classified as an **invertebrate.** About 97 percent of all animal species are invertebrates. Sponges, jellyfish, worms, insects, and clams are examples of invertebrates. Many invertebrates are well protected by their outer coverings. Some have a shell, some have a skeleton on the outside of their body, and others have a spiny outer covering.

Symmetry After determining whether or not a backbone is present, a scientist might look at an animal's symmetry (SIH muh tree). Symmetry is how the body parts of an animal are arranged. Organisms that have no definite shape are called asymmetrical. Most sponges are asymmetrical animals.

Research The classification of an animal can change as new information is learned. Visit the Glencoe Science Web site at **science.glencoe.com** to learn about a recent reclassification of an animal. Communicate to your class what you learn.

Figure 8 Symmetry is a characteristic of all animals.

A Sea urchins have radial symmetry and can sense things in their environment from all directions.

Animals that have body parts arranged in a circle around a center point, the way spokes of a bicycle wheel are arranged, have **radial symmetry.** Hydras, jellyfish, sea urchins like the one in **Figure 8A,** and some sponges have radial symmetry.

Most animals have bilateral symmetry. Look in the mirror. Does your body look about the same on both sides? An animal with **bilateral symmetry** has its body parts arranged in a similar way on both sides. Each half is a mirror image of the other half. In Latin, the word *bilateral* means "two sides." Bilateral animals, like the one in **Figure 8B,** can be divided into right and left halves.

B Most animals have bilateral symmetry like this crayfish. *What type of symmetry do you have?*

After an animal is classified as an invertebrate or a vertebrate and its symmetry is determined, other characteristics are identified that place it in one of the groups of animals with which it has the most characteristics in common. Sometimes a newly discovered animal is different from any existing group, and a new classification group is formed for that animal.

Section Assessment

1. List five characteristics of animals.
2. Explain how herbivores, carnivores, and omnivores are different.
3. Compare invertebrates and vertebrates.
4. Name the three types of symmetry. Give an animal example for each type.
5. **Think Critically** Radial symmetry is found among species that live in water. Why might radial symmetry be an uncommon adaptation of animals that live on land?

Skill Builder Activities

6. **Concept Mapping** Using the information in this section, make an events-chain concept map that shows the steps a scientist might use to classify a new animal. **For more help, refer to the** Science Skill Handbook.
7. **Communicating** Choose an animal you are familiar with. In your Science Journal, describe the adaptations it has for getting food and avoiding predators. **For more help, refer to the** Science Skill Handbook.

SECTION 2 Sponges and Cnidarians

As You Read

What You'll Learn
- **Describe** the characteristics of sponges and cnidarians.
- **Explain** how sponges and cnidarians obtain food and oxygen.
- **Determine** the importance of living coral reefs.

Vocabulary
sessile
hermaphrodite
polyp
medusa
tentacle
stinging cell

Why It's Important
Sponges and cnidarians are important to medical research because they are sources of chemicals that fight disease.

Sponges

When you think of sponges, do you think of the colorful, packaged ones that are used for cleaning or those that live in water? Natural sponges that some humans use are the remains of animals. When alive, they carried on the life processes that all animals do.

Importance of Sponges In their watery environments, sponges play many roles. They interact with many other animals such as worms, shrimp, snails, and sea stars. These animals live on, in, and under sponges. Sponges also are important as a food source for some snails, sea stars, and fish. Certain sponges contain photosynthetic bacteria and protists that provide oxygen and remove wastes for the sponge.

Only about 17 species of sponges are commercially important. Humans have long used the dried and cleaned bodies of some sponges for bathing and cleaning. Most sponges you see today are synthetic sponges or vegetable loofah sponges, but natural sea sponges like those in **Figure 9A** still are available.

Today scientists are finding other uses for sponges. Chemicals made by sponges are being tested and used to make drugs that fight disease-causing bacteria, fungi, and viruses. These chemicals also might be used to treat certain forms of arthritis.

Origin of Sponges Fossil evidence shows that sponges appeared on Earth about 600 million years ago. Because sponges have little in common with other animals, many scientists have concluded that sponges probably evolved separately from all other animals. Sponges living today have many of the same characteristics as their fossilized ancestors.

Figure 9
A Most sponges are found in salt water, but B a few species are found in freshwater.

Characteristics of Sponges

Most of the 5,000 species of sponges are found in warm, shallow salt water near coastlines, although some are found at ocean depths of 8,500 m or more. A few species, like the one in **Figure 9B,** live in freshwater rivers, lakes, and streams. The colors, shapes, and sizes of sponges vary. Saltwater sponges are brilliant red, orange, yellow, or blue, while freshwater sponges are usually a dull brown or green. Some sponges have radial symmetry, but most are asymmetrical. Sponges can be smaller than a marble or larger than a compact car.

Adult sponges live attached to one place unless they are washed away by strong waves or currents. Organisms that remain attached to one place during their lifetimes are called **sessile** (SE sile). They often are found with other sponges in permanent groups called colonies. Early scientists classified sponges as plants because they didn't move. As microscopes were improved, scientists observed that sponges couldn't make their own food, so sponges were reclassified as animals.

Body Structure A sponge's body, like the one in **Figure 10,** is a hollow tube closed at the bottom and open at the top. The sponge has many small openings in its body. These openings are called pores.

Sponges have less complex body organization than other groups of animals. They have no tissues, organs, or organ systems. The body wall has two cell layers made up of several different types of cells. Those that line the inside of the sponge are called collar cells. The beating motion of the collar cells' flagella moves water through the sponge.

Many sponge bodies contain sharp, pointed structures called spicules (SPIH kyewlz). The soft-bodied, natural sponges that some people use for bathing or washing their cars have skeletons of a fibrous material called spongin. Other sponges contain spicules and spongin. Spicules and spongin provide support for a sponge and protection from predators.

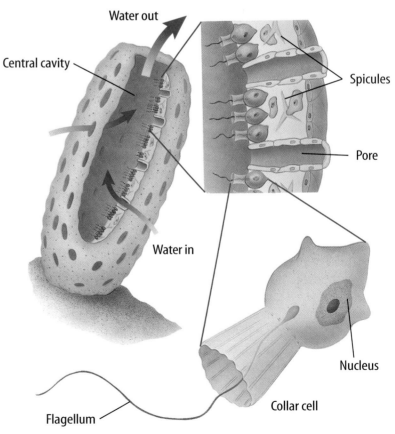

Figure 10
Specialized cells, called collar cells, have flagella that move water through the pores in a sponge. Other cells filter microscopic food from the water as it passes through.

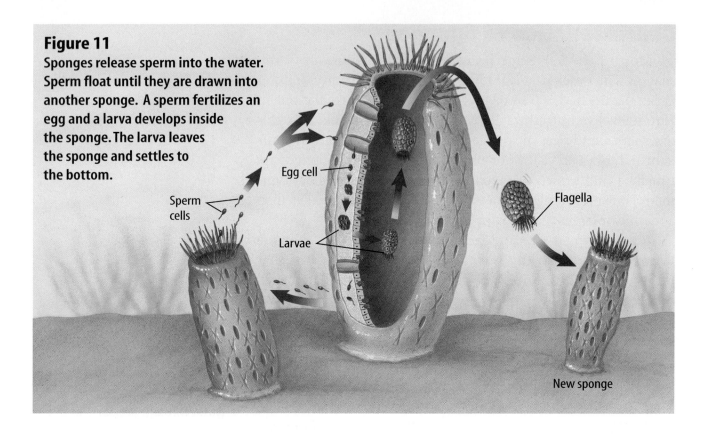

Figure 11
Sponges release sperm into the water. Sperm float until they are drawn into another sponge. A sperm fertilizes an egg and a larva develops inside the sponge. The larva leaves the sponge and settles to the bottom.

Obtaining Food and Oxygen Sponges filter microscopic food particles such as bacteria, algae, protists, and other materials from the water as it is pulled in through their pores. Oxygen also is removed from the water. The filtered water carries away wastes through an opening in the top of the sponge.

✔ **Reading Check** *How do sponges get oxygen?*

Reproduction Sponges can reproduce sexually, as shown in **Figure 11.** Some species of sponges have separate sexes, but most sponge species are **hermaphrodites** (hur MA fruh dites)—animals that produce sperm and eggs in the same body. However, a sponge's sperm cannot fertilize its own eggs. After an egg is released, it might be fertilized and then develop into a larva (plural, *larvae*). The larva usually looks different from the adult form. Sponge larvae have cilia that allow them to swim. After a short time, the larvae settle down on objects where they will remain and grow into adult sponges.

Asexual reproduction occurs by budding or regeneration. A bud forms on a sponge, then drops from the parent sponge to grow on its own. New sponges also can grow by regeneration from small pieces of a sponge. Regeneration occurs when an organism grows new body parts to replace lost or damaged ones. Sponge growers cut sponges into pieces, attach weights to them, and put them back into the ocean to regenerate.

Chemistry INTEGRATION

Spicules of glass sponges are composed of silica. Other sponges have spicules of calcium carbonate. Relate the composition of spicules to the composition of the water in which the sponge lives. Write your answer in your Science Journal.

Figure 18
The planarian is a common freshwater flatworm.

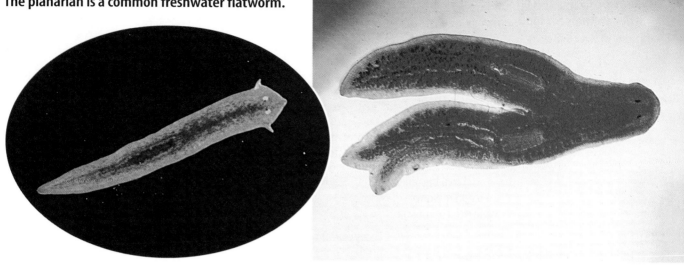

A The planarian's eyespots sense light.

B Planarians can reproduce asexually by splitting, then regenerating the other half.

Planarians reproduce asexually by dividing in two, as shown in **Figure 18B.** A planarian can be cut in two, and each piece will grow into a new worm. They also have the ability to regenerate. Planarians reproduce sexually by producing eggs and sperm. Most are hermaphrodites and exchange sperm with one another. They lay fertilized eggs that hatch in a few weeks.

Flukes All flukes are parasites with complex life cycles that require more than one host. Most flukes reproduce sexually. The male worm deposits sperm in the female worm. She lays the fertilized eggs inside the host. The eggs leave the host in its urine or feces. If the eggs end up in water, they usually infect snails. After they leave the snail, the young worms can burrow into the skin of a new host, such as a human, while he or she is standing or swimming in the water.

Of the many diseases caused by flukes, the most widespread one affecting humans is schistosomiasis (shis tuh soh MI uh sus). It is caused by blood flukes—flatworms that live in the blood, as shown in **Figure 19.** More than 200 million people, mostly in developing countries, are infected with blood flukes. It is estimated that 1 million people die each year because of them. Other types of flukes can infect the lungs, liver, eyes, and other organs of their host.

Figure 19
Female blood flukes deposit their eggs in the blood of their host. The eggs travel through the host and eventually end up in the host's digestive system.

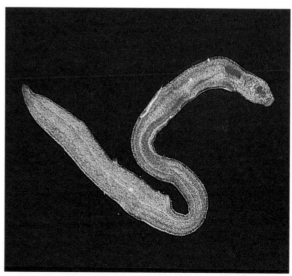

Magnification: 20×

Reading Check *What is the most common disease that is caused by flukes?*

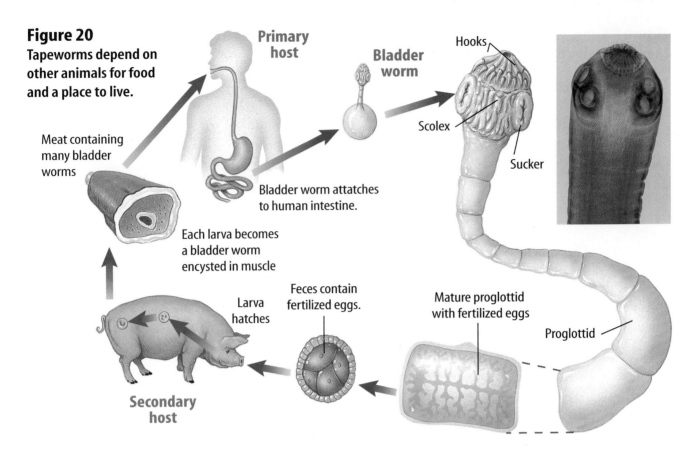

Figure 20 Tapeworms depend on other animals for food and a place to live.

Tapeworms Another type of flatworm is the tapeworm. These worms are parasites. The adult form uses hooks and suckers to attach itself to the intestine of a host organism, as illustrated in **Figure 20.** Dogs, cats, humans, and other animals are hosts for tapeworms. A tapeworm doesn't have a mouth or a digestive system. Instead, the tapeworm absorbs food that is digested by the host from its host's intestine.

A tapeworm grows by producing new body segments immediately behind its head. Its ribbonlike body can grow to be 12 m long. Each body segment has both male and female reproductive organs. The eggs are fertilized by sperm in the same segment. After a segment is filled with fertilized eggs, it breaks off and passes out of the host's body with the host's wastes. If another host eats a fertilized egg, the egg hatches and develops into a new tapeworm.

Origin of Flatworms

Because of the limited fossil evidence, the evolution of flatworms is uncertain. Evidence suggests that they were the first group of animals to evolve bilateral symmetry with senses and nerves in the head region. They also were probably the first group of animals to have a third tissue layer that develops into organs and systems. Some scientists hypothesize that flatworms and cnidarians might have had a common ancestor.

Mini LAB

Observing Planarian Movement

Procedure
1. Use a **dropper** to transfer a **planarian** to a **watch glass.**
2. Add enough **water** so the planarian can move freely.
3. Place the glass under a **stereomicroscope** and observe the planarian.

Analysis
1. Describe how a planarian moves in the water.
2. What body parts appear to be used in movement?
3. Explain why a planarian is a free-living flatworm.

Roundworms

If you own a dog, you've probably had to get medicine from your veterinarian to protect it from heartworms—a type of roundworm. Roundworms are also called the nematodes and make up the largest group of worms. More nematodes live on Earth than any other type of many-celled organism. It is estimated that more than a half million species of roundworms exist. They are found in soil, animals, plants, freshwater, and salt water. Many are parasitic, but most are free-living.

Roundworms are slender and tapered at both ends like the one in **Figure 21**. The body is a tube within a tube, with fluid in between. Most nematode species have male and female worms and reproduce sexually. Unlike the other animals in this chapter, nematodes have two body openings, a mouth, and an anus. The **anus** is an opening at the end of the digestive tract through which wastes leave the body.

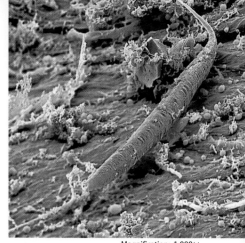

Magnification: 1,000×

Figure 21
Some roundworms infect humans and other animals. Others infect plants, and some are free-living in the soil.

Math Skills Activity

Using Percent to Calculate Number of Species

Example Problem

In a specific deciduous forest ecosystem, 400 different species of animals exist. Scientists estimate that roundworms make up about four percent of these animal species. How many roundworms probably are present in the deciduous forest ecosystem?

Solution

1. *This is what you know:*   There are 400 known species of animals. Roundworms make up four percent of all animal species.

2. *This is what you need to do:*   Change four percent to a decimal.

$$\frac{4}{100} = 0.04$$

Calculate the number of roundworms using the following equation.
(total number of animal species) × (percent of species as a decimal) = number of species

3. *Substitute the known values:*   400 × 0.04 = 16 roundworm species

Practice Problem

Flatworms make up 1.5 percent of all animal species. How many flatworms species probably are present in the forest ecosystem?

For more help, refer to the **Math Skill Handbook**.

NATIONAL GEOGRAPHIC VISUALIZING PARASITIC WORMS

Figure 22

Many diseases are caused by parasitic roundworms and flatworms that take up residence in the human body. Some of these diseases result in diarrhea, weight loss, and fatigue; others, if left untreated, can be fatal. Micrographs of several species of roundworms and flatworms and their magnifications are shown here.

▶ **78× BLOOD FLUKE** These parasites live as larvae in lakes and rivers and penetrate the skin of people wading in the water. After maturing in the liver, the flukes settle in veins in the intestine and bladder, causing schistosomiasis (shis tuh soh MI uh sus), which damages the liver and spleen.

▼ **125× PINWORMS** Typically inhabiting the large intestine, the female pinworm lays her eggs near the host's anus, causing discomfort. The micrograph below shows pinworm eggs on a piece of clear tape.

▲ **6× LIVER FLUKE** Humans and other mammals ingest the larvae of these parasites by eating contaminated plant material. Immature flukes penetrate the intestinal wall and pass via the liver into the bile ducts. There they mature into adults that feed on blood and tissue.

◀ **170× ROUNDWORMS** The roundworms that cause the disease trichinosis (trih kuh NOH sus) are eaten as larvae in undercooked infected meat. They mature in the intestine, then migrate to muscle tissue, where they form painful cysts.

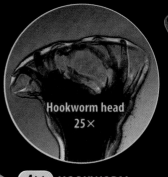
Hookworm head 25×

▶ **4× HOOKWORM** These parasites enter their human hosts as larvae by penetrating the skin of bare feet. From there, they migrate to the lungs and eventually to the intestine, where they mature.

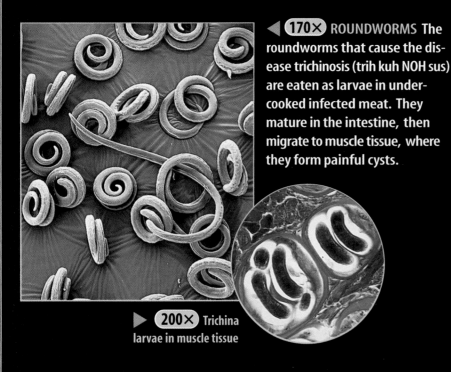

▶ **200× Trichina larvae in muscle tissue**

Origin of Roundworms More than 550 million years ago, roundworms appeared early in animal evolution. They were the first group of animals to have a digestive system with a mouth and an anus. Scientists hypothesize that roundworms are more closely related to arthropods than to vertebrates. However, it is still unclear how roundworms fit into the evolution of animals.

Importance of Roundworms Some roundworms cause diseases in humans, as shown in **Figure 22.** Others are parasites of other animals, like the fish in **Figure 23,** or plants. Some nematodes cause damage to fiber, agricultural products, and food. It is estimated that the worldwide annual amount of nematode damage is in the millions of dollars.

Not all roundworms are a problem for humans, however. In fact, many species are beneficial. Some species of roundworms feed on termites, fleas, ants, beetles, and many other types of insects that cause damage to crops and human property. Some species of beneficial nematodes kill other pests. Research is being done with nematodes that kill deer ticks that cause Lyme disease.

Roundworms also are important because they are essential to the health of soil. They provide nutrients to the soil as they break down organic material. They also help in cycling nutrients such as nitrogen.

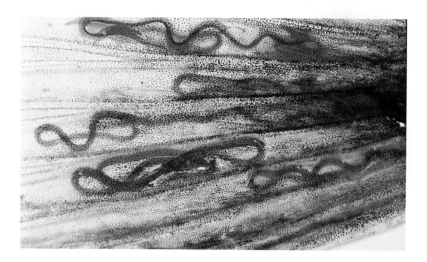

Figure 23
This fish's fin is infected with parasitic roundworms. These roundworms damage the fin, which makes it difficult for the fish to swim and escape from predators.

Section Assessment

1. Compare the body plan of a flatworm to the body plan of a roundworm.
2. Distinguish between a free-living flatworm and a parasitic flatworm.
3. How do tapeworms get energy?
4. What are three roundworms that cause diseases in humans? How can humans prevent infection from each?
5. **Think Critically** Why is a flatworm considered to be more complex than a hydra?

Skill Builder Activities

6. **Concept Mapping** Make an events-chain concept map for tapeworm reproduction. **For more help, refer to the** Science Skill Handbook.
7. **Communicating** In your Science Journal, write a public service announcement for your local radio or television station informing the community about heartworm disease in dogs. Consult a veterinarian for information. **For more help, refer to the** Science Skill Handbook.

Activity: Design Your Own Experiment

Comparing Free-Living and Parasitic Flatworms

Observe free-living and parasitic flatworms to determine how each type of flatworm is adapted to its particular environment.

Recognize the Problem
How are the body parts of flatworms adapted to the environment in which they live?

Form a Hypothesis
State a hypothesis about how free-living and parasitic flatworms are able to live in different environments.

Possible Materials
petri dish with a planarian
compound microscope
prepared slide of a tapeworm
stereomicroscope
light source, such as a lamp
small paintbrush
small piece of liver
dropper
water

Goals
- **Compare and contrast** the body parts and functions of free-living and parasitic flatworms.
- **Observe** how flatworms are adapted to their environments.

Safety Precautions

Using Scientific Methods

Test Your Hypothesis

Plan

1. As a group, make a list of possible ways you might design a procedure to compare and contrast types of flatworms. Your teacher will provide you with information on handling live flatworms.
2. Choose one of the methods you described in step 1. List the steps you will need to take to follow the procedure. Be sure to describe exactly what you will do at each step of the activity.
3. **List** the materials that you will need to complete your experiment.
4. If you need a data table, design one in your Science Journal so it is ready to use when your group begins to collect data.

Do

1. Make sure your teacher approves your plan before you start.

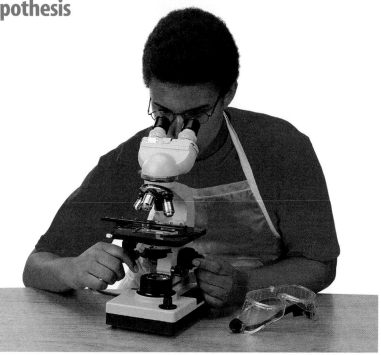

2. Carry out the experiment according to the approved plan.
3. While the experiment is going on, record any observations that you make and complete the data table in your Science Journal.

Analyze Your Data

1. **Explain** how parasitic and free-living flatworms are similar.
2. **Describe** the differences between parasitic and free-living worms.

Draw Conclusions

1. Which body systems are more developed in free-living flatworms?
2. Which body system is more complex in parasitic flatworms?
3. What adaptations allow some flatworms to live as free-living organisms?

Compare and **discuss** your conclusions about body design with other students. **For more help, refer to the** Science Skill Handbook.

ACTIVITY C ◆ 29

TIME SCIENCE AND HISTORY

SCIENCE CAN CHANGE THE COURSE OF HISTORY!

SPONGES

A common household item contains a lot of history

Sponges and baths. They go together like hammer and nails, like burgers and fries. But sponges weren't always used just to scrub people and countertops. True, the ancient Greeks, Romans, and Egyptians did clean themselves with sponges. The Greeks and Romans also mopped floors and wiped tables with sponges. The more artistic Greeks dipped sponges into paint and dabbed it onto their artworks and crafts.

A natural sponge (above) and a synthetic sponge (right) are both used for mopping up spills.

In the past, sponges went to war. Greek and Roman soldiers padded their helmets with soft sponges similar to modern bike or skateboard helmets. This made them more comfortable and helped cushion blows from their enemies' weapons. For Roman soldiers, sponges also served as a lightweight canteen. During a long hot march, soldiers would dunk the sponge in a cool stream or well, soak up the water and squeeze the liquid into their mouth.

Throughout history, people have found sponges an absorbing subject. Pictures of sponges appear in the artwork of the prehistoric civilization of Crete, an island in the Aegean Sea. Medical books in the Middle Ages describe how sponges were used to clean and bathe wounds. The sponge even turns up in Shakespeare. In the play *Hamlet*, one character is described as sponging off the king.

How do you catch a sponge?

No matter how they're used, commercial sponges have been gathered over time from the waters of the Mediterranean, the Caribbean Sea, and off the coast of Florida. In the past, divers carried up sponges from deep water. But today, sponges usually are harvested in shallower water by people in boats. They use a hook attached to a long pole to bring these marine animals to the surface.

On land, the sponges are washed to remove dirt and organisms. Then they are left to dry. Eventually, the sponge's soft tissue rots away. What remains is the skeleton of the sponge. This meshlike material, called spongin, holds lots of liquid.

A diver nears a colorful coral and sponge tunnel off the Bahama Islands.

A close-up detail of a yellow sponge found near the coast of Belize in Central America.

Today, synthetic sponges are used more than natural ones. There's probably no natural sponge in that rectangular sponge you use to clean your bathtub. Natural sponges absorb more water and last longer than the rubber or cellulose kind that most people use. What's their advantage? They're cheaper.

They might be expensive, but natural sponges still have value. In fact, they might one day wipe away some diseases. Medical researchers hypothesize that an enzyme produced by sponges might help in curing cancer. Who says natural sponges are washed up?

CONNECTIONS Brainstorm Work with your classmates to come up with as many sayings and phrases as you can using the word sponge. Use some of them in a story about sponges. Share your stories with the class.

Science Online
For more information, visit science.glencoe.com.

Chapter 1 Study Guide

Reviewing Main Ideas

Section 1 Is it an animal?

1. Animals are many-celled organisms that must find and digest their food.
2. Herbivores eat plants, carnivores eat animals or animal flesh, omnivores eat plants and animals, and detritivores feed on decaying plants and animals.
3. Animals have many ways to escape from predators such as speed, mimicry, protective outer coverings, and camouflage.
4. Invertebrates are animals without backbones. Animals that have backbones are called vertebrates.
5. When body parts are arranged the same way on both sides of the body, it is called bilateral symmetry. If body parts are arranged in a circle around a central point, it is known as radial symmetry. Animals without a specific central point are asymmetrical. *What kind of symmetry does the animal in the photo to the right have?*

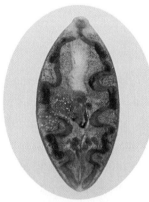

Section 2 Sponges and Cnidarians

1. Adult sponges are sessile and obtain food and oxygen by filtering water and organisms through their pores.
2. Sponges reproduce sexually and asexually.
3. Cnidarians are hollow-bodied animals with radial symmetry. Most have tentacles with stinging cells to obtain food. Jellyfish, hydras, and corals are cnidarians.

4. Coral reefs like the one in the photo below have been deposited by reef-building corals over millions of years. *Which group of animals do corals belong to?*

Section 3 Flatworms and Roundworms

1. Flatworms have bilateral symmetry. Free-living and parasitic forms exist. *Why is a tapeworm considered to be a parasite?*

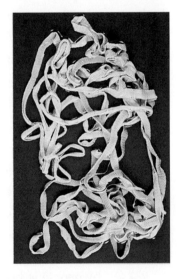

2. Flukes are parasites that have complex life cycles with multiple hosts.
3. Roundworms have a tube-within-a-tube body plan and bilateral symmetry.
4. Flatworms and roundworms have species that cause disease in humans.

After You Read

Add information about other animals to your Organizational Study Fold and then compare and contrast each animal.

Chapter 1 Study Guide

Visualizing Main Ideas

Complete the following concept map.

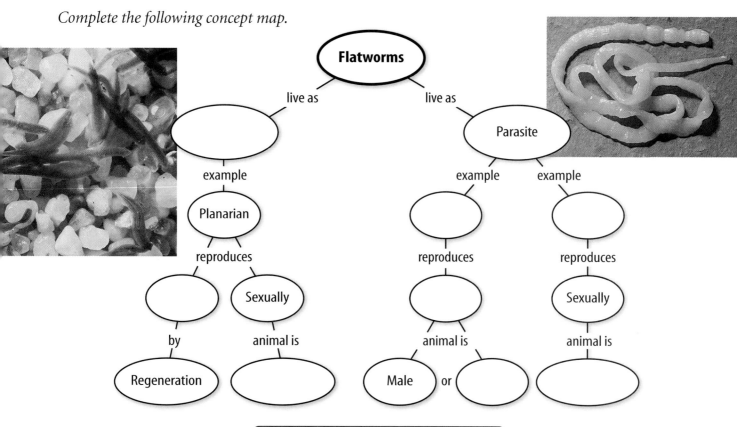

Vocabulary Review

Vocabulary Words

a. anus
b. bilateral symmetry
c. carnivore
d. free-living organism
e. herbivore
f. hermaphrodite
g. invertebrate
h. medusa
i. omnivore
j. polyp
k. radial symmetry
l. sessile
m. stinging cell
n. tentacle
o. vertebrate

Using Vocabulary

Replace the underlined phrases with the correct vocabulary word(s).

1. Jellyfish are <u>animals without backbones</u> and have <u>body parts arranged in a circle around a central point</u>.

2. <u>Animals that eat only other animals</u> eat less often than <u>animals that eat just plants</u>.

3. Most sponges are <u>animals that produce sperm and eggs in one body</u>.

4. Fish, humans, birds, and snakes are <u>animals with backbones</u> and <u>body parts arranged similarly on both sides of the body</u>.

5. Sea anemones are <u>vase shaped</u> and are <u>attached to one place</u>.

Study Tip

Think of other possible ways that you might design an experiment to prove or disprove scientific principles.

Chapter 1 Assessment

Checking Concepts

Choose the word or phrase that best answers the question.

1. Which of the following animals is sessile?
   A) jellyfish C) planarian
   B) roundworm D) sponge

2. What characteristic do all animals have?
   A) digest their food
   B) radial symmetry
   C) free-living
   D) polyp and medusa forms

3. Which term best describes a hydra?
   A) carnivore C) herbivore
   B) filter feeder D) parasite

4. Which animal has a mouth and an anus?
   A) roundworm C) planarian
   B) jellyfish D) tapeworm

5. What characteristic do scientists use to classify sponges?
   A) material that makes up their skeletons
   B) method of obtaining food
   C) reproduction
   D) symmetry

6. Which animal is a cnidarian?
   A) fluke C) jellyfish
   B) heartworm D) sponge

7. Which of the following invertebrate organisms is a hermaphrodite?
   A) fluke C) tapeworm
   B) coral D) roundworm

8. How do sponges reproduce asexually?
   A) budding C) medusae
   B) polyps D) eggs and sperm

9. What is the young organism that the fertilized egg of a sponge develops into?
   A) bud C) medusa
   B) larva D) polyp

10. Which group do roundworms belong to?
    A) cnidarians C) planarians
    B) nematodes D) sponges

Thinking Critically

11. Compare the body organization of a sponge to that of a flatworm.

12. What advantage does being able to reproduce sexually and asexually have for animals like sponges, cnidarians, and flatworms?

13. List the types of food that sponges, hydras, and planarians eat. Explain why each organism eats a different size of particle.

14. Compare and contrast the medusa and polyp body forms of cnidarians.

15. Why do scientists think the medusa stage was the first stage of the cnidarians?

Developing Skills

16. **Concept Mapping** Complete the following concept map about cnidarians.

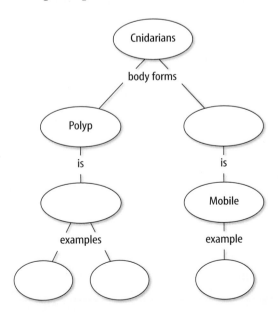

34 ◆ C CHAPTER ASSESSMENT

Chapter 1 Assessment

17. Forming Hypotheses Hypothesize why cooking pork at high temperatures prevents harmful roundworms from developing, if they are present in the uncooked meat.

18. Predicting What can you predict about the life of an organism that has no mouth or digestive system but has suckers and hooks on its head?

19. Interpreting Scientific Illustrations Look at the photograph below. This animal escapes from predators by mimicry. Where in nature might you find the animal in this photo?

20. Identifying and Manipulating Variables and Controls Design an experiment to test the sense of touch in a planarian. Identify variables, constants, and controls.

Performance Assessment

21. Report Research tapeworms and other parasitic worms that live in humans. Find out how they are able to live in the intestines without being digested by the human host. Report your findings to the class.

22. Video Presentation Create a video presentation using computer software or slides to illustrate the variety of sponges and cnidarians found on a coral reef.

TECHNOLOGY

Go to the Glencoe Science Web site at **science.glencoe.com** or use the **Glencoe Science CD-ROM** for additional chapter assessment.

THE PRINCETON REVIEW — Test Practice

Lucas is doing a report about endangered species around the world. He found the following information about endangered animal species in various countries. He placed the information in the table below.

Endangered Species			
Country	Mammal Species	Bird Species	Fish Species
Australia	58	45	37
Indonesia	128	104	60
Thailand	34	45	14
United States	35	50	123
DRC*	38	26	1

*Democratic Republic of Congo

Study the table and answer the following questions.

1. According to the information in the table, the country with the most endangered animals is _____.
   A) the United States
   B) Thailand
   C) Australia
   D) Indonesia

2. According to this information, in which country do endangered mammals account for more than half of the total number of endangered animal species?
   F) United States   H) Thailand
   G) Indonesia   J) DRC

CHAPTER

2 Mollusks, Worms, Arthropods, Echinoderms

What do fiddler crabs running along the beach have in common with pill bugs and millipedes you might discover under a rock in your yard? How many mosquitoes have bitten you? These animals and more than a million other species belong to the largest, most diverse group of animals—the arthropods. In this chapter, you will learn about arthropods, as well as mollusks, worms, and echinoderms.

What do you think?
Science Journal Look at the picture below with a classmate. Discuss what you think this might be. Here's a hint: *It can be found in your house.* Write your answer or best guess in your Science Journal.

If you've ever walked along a beach, especially after a storm, you've probably seen many seashells. They come in many different colors, shapes, and sizes. If you look closely, you will see that some shells have many rings or bands. In the following activity, find out what the bands tell you about the shell and the organism that made it.

Examine a clam's shell

1. Use a hand lens to examine a clam's shell.
2. Count the number of rings or bands on the shell. Count as number one the large, top point called the crown.
3. Compare the distances between the bands of the shell.

Observe

Do other students' shells have the same number of bands? Are all of the bands on your shell the same width? What do you think the bands represent, and why are some wider than others? Record your answers in your Science Journal.

Before You Read

Making an Organizational Study Fold Make the following Foldable to help you organize your thoughts into clear categories about groups of invertebrates.

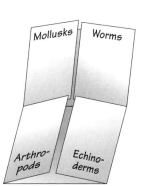

1. Place a sheet of paper in front of you so the long side is at the top. Fold the paper in half from the left side to the right side and then unfold.
2. Fold each side in to the centerfold line to divide the paper into fourths. Fold the paper in half from top to bottom and unfold.
3. Through the top thickness of paper, cut along both of the middle fold lines to form four tabs as shown. Label the tabs *Mollusks, Worms, Arthropods,* and *Echinoderms* as shown.
4. Before you read the chapter, write what you know about each group of invertebrate under the tab. As you read the chapter, add to and correct what you have written.

SECTION 1 Mollusks

As You Read

What You'll Learn
- **Identify** the characteristics of mollusks.
- **Describe** gastropods, bivalves, and cephalopods.
- **Explain** the environmental importance of mollusks.

Vocabulary
mantle
gill
open circulatory system
radula
closed circulatory system

Why It's Important
Mollusks are a food source for many animals. They also filter impurities from the water.

Characteristics of Mollusks

Mollusks (MAH lusks) are soft-bodied invertebrates with bilateral symmetry and usually one or two shells. Their organs are in a fluid-filled cavity. The word *mollusk* comes from the Latin word meaning "soft". Most mollusks live in water, but some live on land. Snails, clams, and squid are examples of mollusks. More than 110,000 species of mollusks have been identified.

Body Plan All mollusks, like the one in **Figure 1,** have a thin layer of tissue called a mantle. The **mantle** covers the body organs, which are located in the visceral (VIH suh rul) mass. Between the soft body and the mantle is a space called the mantle cavity. It contains **gills**—the organs in which carbon dioxide from the mollusk is exchanged for oxygen in the water.

The mantle also secretes the shell or protects the body if the mollusk does not have a shell. The shell is made up of several layers. The inside layer is the smoothest. It is usually the thickest layer because it's added to throughout the life of the mollusk. The inside layer also protects the soft body.

The circulatory system of most mollusks is an open system. In an **open circulatory system,** the heart moves blood through vessels and out into open spaces around the body organs. The body organs are surrounded completely by blood that contains nutrients and oxygen.

Most mollusks have a well-developed head with a mouth and some sensory organs. Some mollusks, such as squid, have tentacles. On the underside of a mollusk is the muscular foot, which is used for movement.

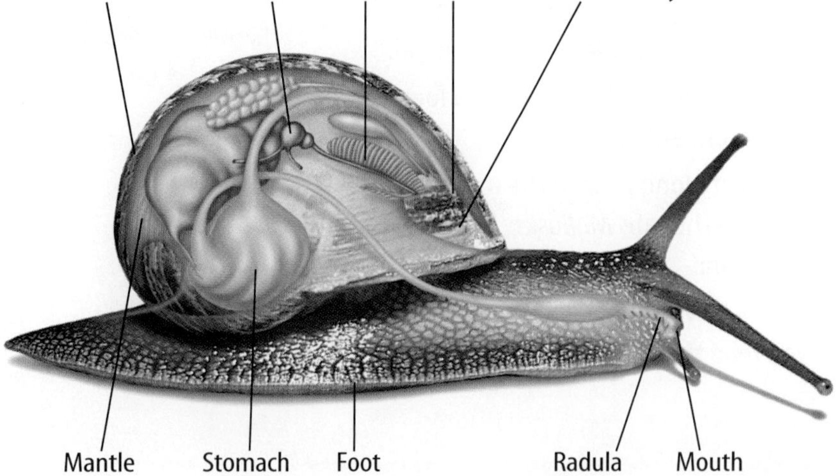

Figure 1
The general mollusk body plan is shown by this snail. Most mollusks have a head, foot, and visceral mass.

(Labels: Shell, Heart, Gill, Anus, Mantle cavity, Mantle, Stomach, Foot, Radula, Mouth)

38 ◆ C CHAPTER 2 Mollusks, Worms, Arthropods, Echinoderms

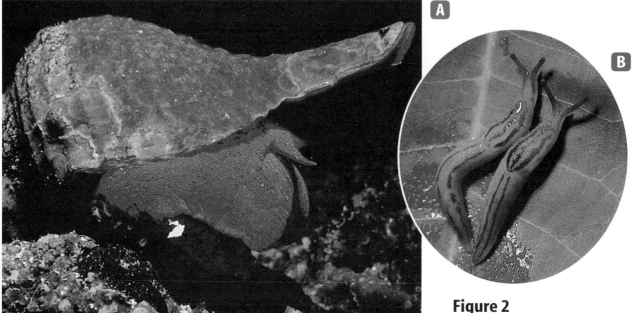

Figure 2
A Conchs, sometimes called marine snails, have a single shell covering their internal organs.
B Garden slugs are mollusks without a shell.

Classification of Mollusks

The first thing scientists look at when they classify mollusks is whether or not the animal has a shell. Mollusks that have shells are then classified by the kind of shell and kind of foot that they have. The three most common groups of mollusks are gastropods, bivalves, and cephalopods.

Gastropods The largest group of mollusks, the gastropods, includes snails, conchs like the one in **Figure 2A,** abalones, whelks, sea slugs, and garden slugs shown in **Figure 2B.** Conchs are sometimes called univalves. Except for slugs, which have no shell, gastropods have a single shell. Many have a pair of tentacles with eyes at the tips. Gastropods use a **radula** (RA juh luh)—a tonguelike organ with rows of teeth—to obtain food. The radula works like a file to scrape and tear food materials. That's why snails are helpful to have in an aquarium—they scrape the algae off the walls and keep the tank clean.

Reading Check *How do gastropods get food?*

Slugs and many snails are adapted to life on land. They move by rhythmic contractions of the muscular foot. Glands in the foot secrete a layer of mucus on which they slide. Slugs and snails are most active at night or on cloudy days when they can avoid the hot sun. Slugs do not have shells but are protected by a layer of mucus instead, so they must live in moist places. Slugs and land snails damage plants as they eat leaves and stems.

Figure 3
Scallops force water between their valves to move away from sea stars and other predators. They can move up to 1 m with each muscular contraction.

Bivalves Mollusks that have a hinged, two-part shell joined by strong muscles are called bivalves. Clams, oysters, and scallops, as shown in **Figure 3,** are bivalve mollusks and are a familiar source of seafood. These animals pull their shells closed by contracting powerful muscles near the hinge. To open their shells, they relax these muscles.

Bivalves are well adapted for living in water. For protection, clams burrow deep into the sand by contracting and relaxing their muscular foot. Mussels and oysters attach themselves with a strong thread or cement to a solid surface. This keeps waves and currents from washing them away. Scallops escape predators by rapidly opening and closing their shells. As water is forced out, the scallop moves rapidly in the opposite direction.

Cephalopods The most specialized and complex mollusks are the cephalopods (SE fuh luh pawdz). Squid, octopuses, cuttlefish, and chambered nautiluses belong to this group. The word *cephalopod* means "head-footed" and describes the body structure of these invertebrates. Cephalopods, like the cuttlefish in **Figure 4,** have a large, well-developed head. Their foot is divided into many tentacles with strong suction cups or hooks for capturing prey. All cephalopods are predators. They feed on fish, crustaceans, worms, and other mollusks.

Squid and octopuses have a well-developed nervous system and large eyes similar to human eyes. Unlike other mollusks, cephalopods have closed circulatory systems. In a **closed circulatory system,** blood containing food and oxygen moves through the body in a series of closed vessels, just as your blood moves through your blood vessels.

Figure 4
Most cephalopods, like this cuttlefish, have an internal shell.

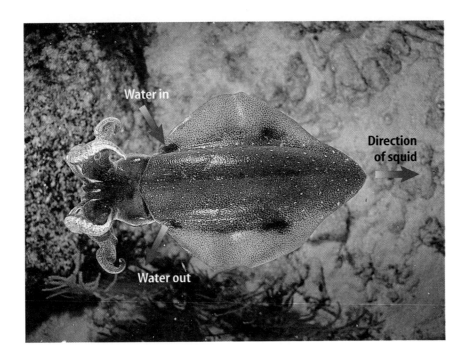

Figure 5
Squid and other cephalopods use jet propulsion to move quickly away from predators.

Cephalopod Propulsion All cephalopods live in oceans and are adapted for swimming. Squid and other cephalopods have a water-filled cavity between an outer muscular covering and its internal organs. When the cephalopod tightens its muscular covering, water is forced out through an opening near the head, as shown in **Figure 5.** The jet of water propels the cephalopod backwards, and it moves away quickly. According to Newton's third law of motion, when one object exerts a force on a second object, the second object exerts a force on the first that is equal and opposite in direction. The movement of cephalopods is an example of this law. Muscles exert force on water under the mantle. Water being forced out exerts a force that results in movement backwards.

A squid can propel itself at speeds of more than 60 m/s using this jet propulsion and can briefly outdistance all but whales, dolphins, and the fastest fish. A squid even can jump out of the water and reach heights of almost 5 m above the ocean's surface. It then can travel through the air as far as 15 m. However, squid can maintain their top speed for just a few pulses. Octopuses also can swim by jet propulsion, but they usually use their tentacles to creep more slowly over the ocean floor.

Origin of Mollusks Some species of mollusks, such as the chambered nautilus, have changed little from their ancestors. Mollusk fossils date back more than 500 million years. Many species of mollusks became extinct about 66 million years ago. Today's mollusks are descendants of ancient mollusks.

By about 225 million years ago, many mollusks had become extinct. Find out why so many mollusks died out. What were the major physical events of the time that could have contributed to changing the environment? Write your answers in your Science Journal.

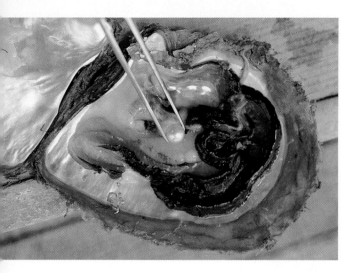

Figure 6
A pearl starts as an irritant—a grain of sand or a parasite—to an oyster. The oyster coats the irritant with a material that forms smooth, hard layers. It can take years for a pearl to form. Culturing pearls is a commercial industry in some countries.

Value of Mollusks

Mollusks have many uses. They are food for fish, sea stars, birds, and humans. Many people make their living raising or collecting mollusks to sell for food. Other invertebrates, such as hermit crabs, use empty mollusk shells as shelter. Many mollusk shells are used for jewelry and decoration. Pearls are produced by several species of mollusks, but most are made by mollusks called pearl oysters, shown in **Figure 6.** Mollusk shells also provide information about the conditions in an ecosystem, including the source and distribution of water pollutants. The internal shell of a cuttlefish is the cuttlebone, which is used in birdcages to provide birds with calcium. Squid and octopuses are able to learn tasks, so scientists are studying their nervous systems to understand how learning takes place and how memory works.

Even though mollusks are beneficial in many ways, they also can cause problems for humans. Land slugs and snails damage plants. Certain species of snails are hosts of parasites that infect humans. Shipworms, a type of bivalve, make holes in submerged wood of docks and boats, causing millions of dollars in damage each year. Because clams, oysters, and other mollusks are filter feeders, bacteria, viruses, and toxic protists from the water can become trapped in the animals. Eating these infected mollusks can result in sickness or even death.

Section Assessment

1. What are the characteristics used to classify mollusks?
2. Name the three groups of mollusks. Identify a mollusk from each group, and explain why it is in that group.
3. Explain how a squid and other cephalopods can move so rapidly.
4. Describe some positive and negative ways that mollusks affect humans.
5. **Think Critically** Why is it unlikely that you would find garden slugs and land snails in a desert?

Skill Builder Activities

6. **Interpreting Scientific Illustrations** Observe the images of gastropods and bivalves in this section. Infer how bivalves are not adapted to life on land and gastropods are. **For more help, refer to the Science Skill Handbook.**

7. **Using a Word Processor** Make a data table that compares and contrasts the following for gastropods, bivalves, and cephalopods: *methods of obtaining food, movement, circulation,* and *habitat.* **For more help, refer to the Technology Skill Handbook.**

SECTION 2 Segmented Worms

Segmented Worm Characteristics

The worms you see crawling across sidewalks after a rain and those used for fishing are called annelids (A nul udz). The word *annelid* means "little rings" and describes the bodies of these worms. They have tube-shaped bodies that are divided into many segments.

Have you ever watched a robin try to pull an earthworm out of the ground or tried it yourself? Why don't they slip out of the soil easily? On the outside of each body segment are bristlelike structures called **setae** (SEE tee). Segmented worms use their setae to hold on to the soil and to move. Segmented worms also have bilateral symmetry, a body cavity that holds the organs, and two body openings—a mouth and an anus. Annelids can be found in freshwater, salt water, and moist soil. Earthworms, like the one in **Figure 7,** marine worms, and leeches are examples of annelids.

Reading Check *What is the function of setae?*

Earthworm Body Systems

The most well-known annelids are earthworms. They have a definite anterior, or front end, and a posterior, or back end. Earthworms have more than 100 body segments. The segments can be seen on the outside and the inside of the body cavity. Each body segment, except for the first and last segments, has four pairs of setae. Earthworms move by using their setae and two sets of muscles in the body wall. One set of muscles runs the length of the body, and the other set circles the body. When an earthworm contracts its long muscles, it causes some of the segments to bunch up and the setae to stick out. This anchors the worm to the soil. When the circular muscles contract, the setae are pulled in and the worm can move forward.

As You Read

What You'll Learn
- **Identify** the characteristics of segmented worms.
- **Describe** the structures of an earthworm and how it takes in and digests food.
- **Explain** the importance of segmented worms.

Vocabulary
setae
crop
gizzard

Why It's Important
Earthworms condition and aerate the soil, which helps increase crop yields.

Figure 7
One species of earthworm that lives in Australia can grow to be 3.3 m long.

Digestion and Excretion As an earthworm burrows through the soil, it takes soil into its mouth. Earthworms get energy from the bits of leaves and other organic matter found in the soil. The soil ingested by an earthworm moves to the **crop,** which is a sac used for storage. Behind the crop is a muscular structure called the **gizzard,** which grinds the soil and the bits of organic matter. This ground material passes to the intestine, where the organic matter is broken down and the nutrients are absorbed by the blood. Wastes leave the worm through the anus. When earthworms take in soil, they provide spaces for air and water to flow through it and mix the soil. Their wastes pile up at the openings to their burrows. These piles are called castings. Castings, like those in **Figure 8,** help fertilize the soil.

Figure 8
Earthworm castings—also called vermicompost—are used as an organic fertilizer in gardens.

Circulation and Respiration Earthworms have a closed circulatory system, as shown in **Figure 9.** Two blood vessels along the top of the body and one along the bottom of the body meet in the front end of the earthworm. There, they connect to heartlike structures called aortic arches, which pump blood through the body. Smaller vessels go into each body segment.

Earthworms don't have gills or lungs. Oxygen and carbon dioxide are exchanged through their skin, which is covered with a thin film of watery mucus. It's important never to touch earthworms with dry hands or remove their thin mucous layer, because they could suffocate. But as you can tell after a rainstorm, earthworms don't survive in puddles of water either.

Figure 9
An earthworm has five aortic arches that pump blood throughout its body.

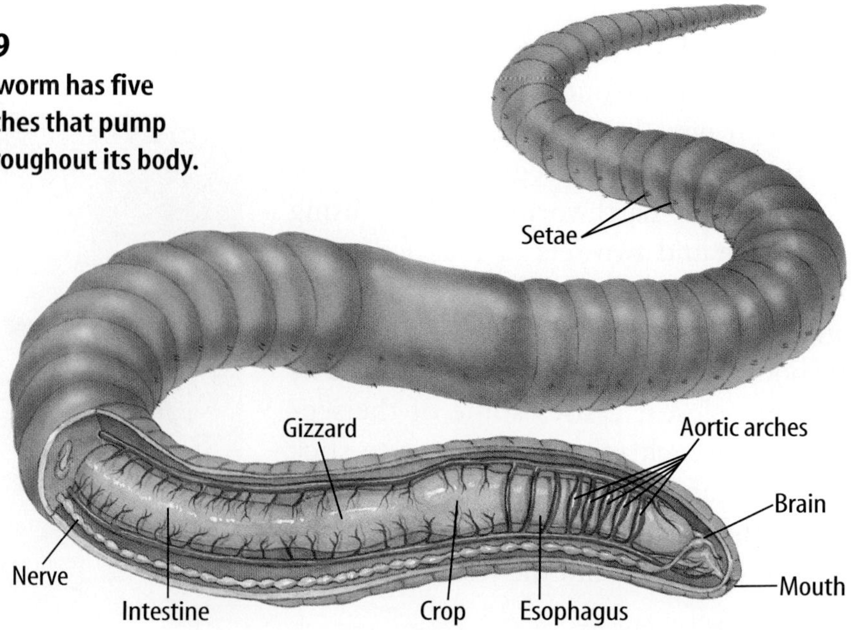

Nerve Response and Reproduction Earthworms have a small brain in their front segment. Nerves in each segment join to form a main nerve cord that connects to the brain. Earthworms respond to light, temperature, and moisture.

Earthworms are hermaphrodites (hur MAF ruh dites)—meaning they produce sperm and eggs in the same body. Even though each worm has male and female reproductive structures, an individual worm can't fertilize its own eggs. Instead, it has to receive sperm from another earthworm in order to reproduce.

Marine Worms

More than 8,000 species of marine worms, or polychaetes, (PAH lee keets) exist, which is more than any other kind of annelid. Marine worms float, burrow, build structures, or walk along the ocean floor. Some polychaetes even produce their own light. Others, like the ice worms in **Figure 10,** are able to live 540 m deep. Polychaetes, like earthworms, have segments with setae. However, the setae occur in bundles on these worms. The word *polychaete* means "many bristles."

Sessile, bottom-dwelling polychaetes, such as the fan worm shown in **Figure 11A,** have specialized tentacles that are used for exchanging oxygen and carbon dioxide and gathering food. Some marine worms build tubes around their bodies. When these worms are startled, they retreat into their tubes. Free-moving polychaetes, such as the bristleworm shown in **Figure 11B,** have a head with eyes; a tail; and parapodia (pur uh POH dee uh)—paired fleshy outgrowths on their segments. The parapodia help in feeding and locomotion.

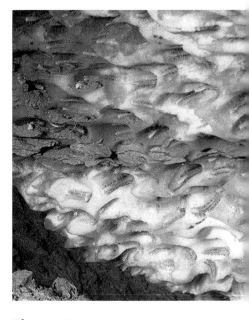

Figure 10
Ice worms, a type of marine polychaete, were discovered first in 1997 living 540 m deep in the Gulf of Mexico.

Figure 11
A This Christmas tree worm filter microorganisms from the water to eat. **B** Some free-swimming polychaetes swim backwards and forwards, so some have eyes at both ends of their body.

SECTION 3 Arthropods

As You Read

What You'll Learn
- **Determine** the characteristics that are used to classify arthropods.
- **Explain** how the structure of the exoskeleton relates to its function.
- **Distinguish** between complete and incomplete metamorphosis.

Vocabulary
appendage
exoskeleton
molting
spiracle
metamorphosis

Why It's Important
Arthropods, such as those that carry diseases and eat crops, affect your life every day.

Characteristics of Arthropods

There are more than a million different species of arthropods, (AR thruh pahdz) making them the largest group of animals. The word *arthropoda* means "jointed foot." The jointed **appendages** of arthropods can include legs, antennae, claws, and pincers. Arthropod appendages are adapted for moving about, capturing prey, feeding, mating, and sensing their environment. Arthropods also have bilateral symmetry, segmented bodies, an exoskeleton, a body cavity, a digestive system with two openings, and a nervous system. Most arthropod species have separate sexes and reproduce sexually. Arthropods are adapted to living in almost every environment. They vary in size from microscopic dust mites to the large, Japanese spider crab, shown in **Figure 14**.

Segmented Bodies The bodies of arthropods are divided into segments similar to those of segmented worms. Some arthropods have many segments, but others have segments that are fused together to form body regions, such as those of insects, spiders, and crabs.

Exoskeletons All arthropods have a hard outer covering called an **exoskeleton**. It covers, supports, and protects the internal body and provides places for muscles to attach. In many land-dwelling arthropods, for example insects, the exoskeleton has a waxy layer that reduces water loss from the animal.

An exoskeleton cannot grow as the animal grows. From time to time, it is shed and replaced by a new one in a process called **molting**. While the animals are molting, they are not well protected from predators because the new exoskeleton is soft. Before the new exoskeleton hardens, the animal swallows air or water to increase its exoskeleton's size. This way the new exoskeleton allows room for growth.

Figure 14
The Japanese spider crab has legs that can span more than 3m.

Insects

More species of insects exist than all other animal groups combined. More than 700,000 species of insects have been classified, and scientists identify more each year. Insects have three body regions—a head, a thorax, and an abdomen, as shown in **Figure 15**. However, it is almost impossible on some insects to see where one region stops and the next one begins.

Figure 15
One of the largest types of ants is the carpenter ant. Like all insects, it has a head, thorax, and abdomen.

Head An insect's head has a pair of antennae, eyes, and a mouth. The antennae are used for touch and smell. The eyes are simple or compound. Simple eyes detect light and darkness. Compound eyes, like those in **Figure 16,** contain many lenses and can detect colors and movement. The mouthparts of insects vary, depending on what the insect eats.

Thorax Three pairs of legs and one or two pairs of wings, if present, are attached to the thorax. Some insects, such as silverfish and fleas, don't have wings, and other insects have wings only for part of their lives. Insects are the only invertebrate animals that can fly. Flying allows insects to find places to live, food sources, and mates. Flight also helps them escape from their predators.

Reading Check *How does flight benefit insects?*

Abdomen The abdomen has neither wings nor legs but it is where the reproductive structures are found. Females lay thousands of eggs, but only a fraction of the eggs develop into adults. Think about how overproduction of eggs might ensure that each insect species will continue.

Insects have an open circulatory system that carries digested food to cells and removes wastes. However, insect blood does not carry oxygen because it does not have hemoglobin. Instead, insects have openings called **spiracles** (SPIHR ih kulz) on the abdomen and thorax through which air enters and waste gases leave the insect's body.

Figure 16
Each compound eye is made up of small lenses that fit together. Each lens sees a part of the picture to make up the whole scene. Insects can't focus their eyes. Their eyes are always open and can detect movements.

Magnification: 5×

Arachnids

Spiders, scorpions, mites, and ticks are examples of arachnids (uh RAK nudz). They have two body regions—a head-chest region called the cephalothorax (se fuh luh THOR aks) and an abdomen. Arachnids have four pairs of legs but no antennae. Many arachnids are adapted to kill prey with poison glands, stingers, or fangs. Others are parasites.

Scorpions Arachnids that have a sharp, poison-filled stinger at the end of their abdomen are called scorpions. The venom from the stinger paralyzes the prey. Unlike other arachnids, scorpions have a pair of well-developed appendages—pincers—with which they grab their prey. The sting of a scorpion is painful and can be fatal to humans.

Math Skills Activity

Calculating Percent of Elasticity

Example Problem

A strand of spider's silk can be stretched from 65 cm to 85 cm before it loses its elasticity—the ability to snap back to its original length. Calculate the percent of elasticity of spider's silk.

Solution

1. *This is what you know:* original length of silk strand = 65 cm
   stretched length of silk strand = 85 cm

2. *This is what you need to find:* percent of elasticity

3. *This is what you must do:* Find the difference between the stretched and original length. 85 cm − 65 cm = 20 cm

4. *Use the following equation to make your calculations:*
$$\frac{\text{difference in length}}{\text{original length}} \times 100 = \% \text{ of elasticity}$$

$$\frac{20 \text{ cm}}{65 \text{ cm}} \times 100 = 30.7 \% \text{ of elasticity}$$

Practice Problem

A 40-cm strand of nylon can be stretched to a length of 46.5 cm before losing its elasticity. Calculate the percent of elasticity for nylon and compare it to that of spider's silk.

For more help, refer to the Math Skill Handbook.

Spiders Because spiders can't chew their food, they release enzymes into their prey that help digest it. The spider then sucks the predigested liquid into its mouth.

Oxygen and carbon dioxide are exchanged in book lungs, illustrated in **Figure 19.** Openings on the abdomen allow these gases to move into and out of the book lungs.

Mites and Ticks Most mites are animal or plant parasites. However, some are not parasites, like the mites that live in the follicles of human eyelashes. Most mites are so small that they look like tiny specs to the unaided eye. All ticks are animal parasites. Ticks attach to their host's skin and remove blood from their hosts through specialized mouthparts. Ticks often carry bacteria and viruses that cause disease in humans and other animals. Diseases carried by ticks include Lyme disease and Rocky Mountain spotted fever.

Centipedes and Millipedes

Two groups of arthropods—centipedes and millipedes—have long bodies with many segments, exoskeletons, jointed legs, antennae, and simple eyes. They can be found in damp environments, including in woodpiles, under vegetation, and in basements. Centipedes and millipedes reproduce sexually. They make nests for their eggs and stay with them until the eggs hatch.

Compare the centipede and millipede in **Figure 20.** How many pairs of legs does the centipede have per segment? How many pairs of legs does the millipede have per segment? Centipedes hunt for their prey, which includes snails, slugs, and worms. They have a pair of poison claws that they use to inject venom into their prey. Their pinches are painful to humans but usually aren't fatal. Millipedes feed on plants and decaying material and often are found under the damp plant material.

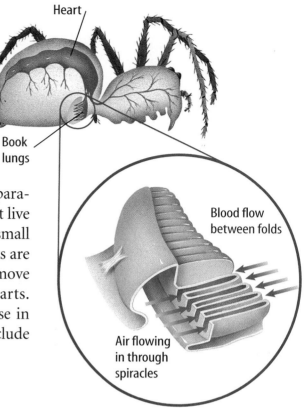

Figure 19
Air circulates between the moist folds of the book lungs bringing oxygen to the blood.

Figure 20
A Centipedes are predators—they capture and eat other animals. **B** Millipedes eat plants or decaying plant material.

NATIONAL GEOGRAPHIC VISUALIZING ARTHROPOD DIVERSITY

Figure 21

Some 600 million years ago, the first arthropods lived in Earth's ancient seas. Today, they inhabit nearly every environment on Earth. Arthropods are the most abundant and diverse group of animals on Earth. They range in size from nearly microscopic mites to spindly, giant Japanese spider crabs with legs spanning more than 3 m.

▲ **LOBSTER** Like crabs, lobsters are crustaceans that belong to the group called Decapoda, which means "ten legs." It's the lobster's tail, however, that interests most seafood lovers.

◀ **GRASS SPIDER** Grass spiders spin fine, nearly invisible webs just above the ground.

◀ **GOOSENECK BARNACLE** Gooseneck barnacles typically live attached to objects that float in the ocean. They use their long, feathery setae to strain tiny bits of food from the water.

▼ **MONARCH BUTTERFLY** Monarchs are a common sight in much of the United States during the summer. In fall, they migrate south to warmer climates.

◀ **HISSING COCKROACH** Most cockroaches are considered to be pests by humans, but hissing cockroaches, such as this one, are sometimes kept as pets.

▶ **HORSESHOE CRAB** Contrary to their name, horseshoe crabs are not crustaceans. They are more closely related to spiders than to crabs.

▶ **CENTIPEDE** One pair of legs per segment distinguishes a centipede from a millipede, which has two pairs of legs per body segment.

Crustaceans

Crabs, crayfish, shrimp, barnacles, pill bugs, and water fleas are crustaceans. Crustaceans and other arthropods are shown in **Figure 21.** Crustaceans have one or two pairs of antennae and mandibles, which are used for crushing food. Most crustaceans live in water, but some, like the pill bugs shown in **Figure 22A,** live in moist environments on land. Pill bugs are common in gardens and around house foundations. They are harmless to humans.

Crustaceans, like the blue crab shown in **Figure 22B,** have five pairs of legs. The first pair of legs are claws that catch and hold food. The other four pairs are walking legs. They also have five pairs of appendages on the abdomen called swimmerets. They help the crustacean move and are used in reproduction. In addition, the swimmerets force water over the feathery gills where the oxygen and carbon dioxide are exchanged. If a crustacean loses an appendage, it will grow back, or regenerate.

Figure 22
A Pill bugs—also called roly polys—are crustaceans that live on land. *How are they similar to centipedes and millipedes?* **B** The segments in some crustaceans aren't obvious because they are covered by a shieldlike structure.

Value of Arthropods

Arthropods play several roles in the environment. They are a source of food for many animals, including humans. Some humans consider shrimp, crab, crayfish, and lobster as food delicacies. In Africa and Asia, many people eat insect larvae and insects such as grasshoppers, termites, and ants, which are excellent sources of protein.

Agriculture would be impossible without bees, butterflies, moths, and flies that pollinate crops. Bees manufacture honey, and silkworms produce silk. Many insects and spiders are predators of harmful animal species, such as stableflies. Useful chemicals are obtained from some arthropods. For example, bee venom is used to treat rheumatic arthritis.

Not all arthropods are useful to humans. Almost every cultivated crop has some insect pest that feeds on it. Many arthropods—mosquitoes, tsetse flies, fleas, and ticks—carry human and other animal diseases. In addition, weevils, cockroaches, carpenter ants, clothes moths, termites, and carpet beetles destroy food, clothing, and property.

Insects are an important part of the ecological communities in which humans live. Removing all of the insects would cause more harm than good.

Controlling Insects One common way to control problem insects is by insecticides. However, many insecticides kill helpful insects as well as harmful ones. Because of their rapid life cycles, many insects have developed resistance to insecticides. Another problem is that many toxic substances that have been used to kill insects remain in the environment and accumulate in the bodies of animals that eat them. As other animals eat the contaminated animals, the insecticides can find their way into human food. Humans also are harmed by these toxins.

Several types of biological controls have been developed and are being tested. Different types of bacteria, fungi, and viruses are being used to control some insect pests. Natural predators and parasites of insect pests have been somewhat successful in controlling certain pests. Other biological controls involve using males that can't reproduce or naturally occurring chemicals that interfere with the reproduction or behavior of insect pests.

Origin of Arthropods Because of their hard body parts, arthropod fossils like the one in **Figure 23** are among the oldest and best-preserved fossils of multicellular animals. Some arthropod fossils are more than 500 million years old. Recall that earthworms and leeches have individual body segments. Because of this, scientists hypothesize that arthropods probably evolved from an ancestor of segmented worms. Over time, groups of body segments fused and became adapted for locomotion, feeding, and sensing the environment. The hard exoskeleton and walking legs allowed arthropods to be among the first animals to live successfully on land.

Figure 23
More than 15,000 species of trilobites have been classified. They are one of the most recognized types of fossils.

Section Assessment

1. What are three characteristics of all arthropods?
2. What are the advantages and disadvantages of an exoskeleton?
3. Compare and contrast insects with arachnids.
4. Which stages of complete and incomplete metamorphosis are different?
5. **Think Critically** Choose an insect you are familiar with and explain how it is adapted to its environment.

Skill Builder Activities

6. **Concept Mapping** Make an events-chain concept map of complete metamorphosis and one of incomplete metamorphosis. **For more help, refer to the Science Skill Handbook.**
7. **Making and Using Graphs** Of the major arthropod groups, 88% are insects, 7% are arachnids, 3% are crustaceans, 1% are centipedes and millipedes, and all others make up 1%. Show these data in a circle graph. **For more help, refer to the Science Skill Handbook.**

Activity

Observing a Crayfish

A crayfish has a segmented body and a fused head and thorax. It has a snout and eyes on movable eyestalks. Most crayfish have pincers. What are the pincers for?

What You'll Investigate
How does a crayfish use its appendages?

Materials
crayfish in a small aquarium
uncooked ground beef
stirrer

Goals
- **Observe** a crayfish.
- **Determine** the function of pincers.

Safety Precautions
WARNING: Use care when working with live animals.

Procedure

1. Copy the data table and use it to record all of your observations during this activity.

Crayfish Observations		
Body Region	Number of Appendages	Function
Head		
Thorax		
Abdomen		

2. Your teacher will provide you with a crayfish in an aquarium. Leave the crayfish in the aquarium while you do the activity. Draw your crayfish.

3. Gently touch the crayfish with the stirrer. How does the body feel?
4. **Observe** how the crayfish moves in the water.
5. **Observe** the compound eyes. On which body region are they located?
6. Drop a small piece of ground beef into the aquarium. Observe the crayfish's reaction. Wash your hands.
7. Return the aquarium to its proper place.

Conclude and Apply

1. **Infer** how the location of the eyes is an advantage for the crayfish.
2. How does the structure of the pincers aid in getting food?
3. What can you infer about the exoskeleton and protection?

Communicating Your Data

Compare your observations with those of other students in your class. **For more help, refer to the** Science Skill Handbook.

SECTION 4 Echinoderms

As You Read

What You'll Learn
- **List** the characteristics of echinoderms.
- **Explain** how sea stars obtain and digest food.
- **Discuss** the importance of echinoderms.

Vocabulary
water-vascular system
tube feet

Why It's Important
Echinoderms are a group of animals that affect oceans and coastal areas.

Echinoderm Characteristics

Echinoderms are found in oceans all over the world. The term *echinoderm* is from the Greek words *echinos* meaning "spiny" and *derma* meaning "skin." Echinoderms have a hard endoskeleton covered by a thin, bumpy or spiny epidermis. They are radially symmetrical, which allows them to sense food, predators, and other things in their environment from all directions.

All echinoderms have a mouth, stomach, and intestines. They feed on a variety of plants and animals. For example, sea stars feed on worms and mollusks, and sea urchins feed on algae. Others feed on dead and decaying matter called detritus (de TRI tus) found on the ocean floor.

Echinoderms have no head or brain, but they do have a nerve ring that surrounds the mouth. They also have cells that respond to light and touch.

Water-Vascular System A characteristic unique to echinoderms is their water vascular system. It allows them to move, exchange carbon dioxide and oxygen, capture food, and release wastes. The **water-vascular system,** as shown in **Figure 24,** is a network of water-filled canals with thousands of tube feet connected to it. **Tube feet** are hollow, thin-walled tubes that each end in a suction cup. As the pressure in the tube feet changes, the animal is able to move along by pushing out and pulling in its tube feet.

Figure 24
Sea stars alternately extend and withdraw their tube feet, enabling them to move.

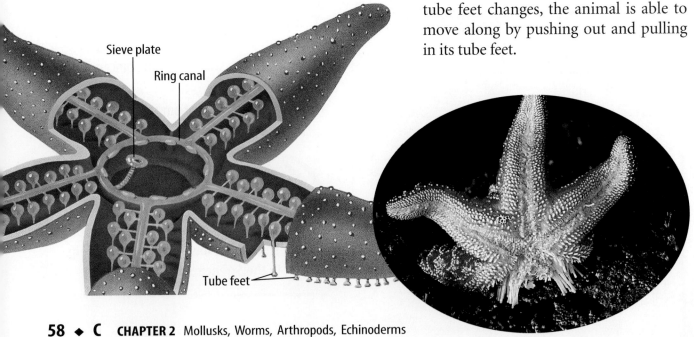

Types of Echinoderms

Approximately 6,000 species of echinoderms (ih KI nuh durmz) are living today. Of those, more than one-third are sea stars. The other groups include brittle stars, sea urchins, sand dollars, and sea cucumbers.

Sea Stars Echinoderms with at least five arms arranged around a central point are called sea stars. The arms are lined with thousands of tube feet. Sea stars use their tube feet to open the shells of their prey. When the shell is open slightly, the sea star pushes its stomach through its mouth and into its prey. The sea star's stomach surrounds the soft body of its prey and secretes enzymes that help digest it. When the meal is over, the sea star pulls its stomach back into its own body.

 *What is unusual about the way that sea stars eat their prey?*

Sea stars reproduce sexually when females release eggs and males release sperm into the water. Females can produce 200,000 eggs in one season.

Sea stars also can repair themselves by regeneration. If a sea star loses an arm, it can grow a new one. If enough of the center disk is left attached to a severed arm, a whole new sea star can grow from that arm.

TRY AT HOME Mini LAB

Modeling the Strength of Tube Feet

Procedure
1. Hold your arm straight out, palm up.
2. Place a heavy **book** on your hand.
3. Have your partner time how long you can hold your arm up with the book on it.

Analysis
1. Describe how your arm feels after a few minutes.
2. If the book models the sea star and your arm models the clam, infer how a sea star successfully overcomes a clam to obtain food.

Brittle Stars Like the one in **Figure 25**, brittle stars have fragile, slender, branched arms that break off easily. This adaptation helps a brittle star survive attacks by predators. While the predator is eating a broken arm, the brittle star escapes. Brittle stars quickly regenerate lost parts. They live hidden under rocks or in litter on the ocean floor. Brittle stars use their flexible arms for movement instead of their tube feet. Their tube feet are used to move particles of food into their mouth.

Figure 25
A brittle star's arms are so flexible that they wave back and forth in the ocean currents. They are called brittle stars because their arms break off easily if they are grabbed by a predator.

Figure 26
Like all echinoderms, sand dollars and sea urchins are radially symmetrical.

A Sand dollars live on ocean floors where they can burrow into the sand.

B Sea urchins use tube feet and their spines to move around on the bottom of the ocean.

Sea Urchins and Sand Dollars Another group of echinoderms includes sea urchins, sea biscuits, and sand dollars. They are disk- or globe-shaped animals covered with spines. They do not have arms, but sand dollars have a five-pointed pattern on their surface. Living sand dollars, like those in **Figure 26A,** are covered with stiff, hairlike spines. Sea urchins, like those in **Figure 26B,** have long, pointed spines that protect them from predators. Some have sacs near the end of the spines that contain poisonous fluid that is injected into predators. The spines also help in movement and burrowing. Sea urchins have five toothlike structures around their mouth.

Sea Cucumbers The animal shown in **Figure 27** is a sea cucumber. Sea cucumbers are soft-bodied echinoderms that have a leathery covering. They have tentacles around their mouth and rows of tube feet on their upper and lower surfaces. When threatened, sea cucumbers may expel their internal organs. These organs regenerate in a few weeks. Some sea cucumbers eat detritus, and others eat plankton.

SCIENCE Online

Research Visit the Glencoe Science Web site at **science.glencoe.com** for information about how echinoderms are used by humans. Communicate to your class what you learn.

Figure 27
Sea cucumbers have short tube feet, which they use to move around.

Value of Echinoderms

Echinoderms are important to the marine environment because they feed on dead organisms and help recycle materials. Sea urchins control the growth of algae in coastal areas. Sea urchin eggs and sea cucumbers are used for food in some places. Many echinoderms are used in research and some might be possible sources of medicines. Sea stars are important predators that control populations of other animals. However, because sea stars feed on oysters and clams, they also destroy millions of dollars' worth of mollusks each year.

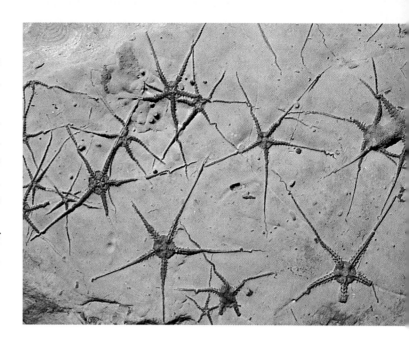

Figure 28
Ophiopinna elegans was a brittle star that lived about 165 million years ago.

Origin of Echinoderms Like the example in **Figure 28**, a good fossil record exists for echinoderms. Echinoderms date back more than 400 million years. The earliest echinoderms might have had bilateral symmetry as adults and may have been attached to the ocean floor by stalks. Many larval forms of modern echinoderms are bilaterally symmetrical.

Scientists hypothesize that echinoderms more closely resemble animals with backbones than any other group of invertebrates. This is because echinoderms have complex body systems and an embryo that develops the same way that the embryos of animals with backbones develop.

Section 4 Assessment

1. What characteristics do all echinoderms have in common?
2. How do echinoderms move and get their food?
3. How are sea urchins beneficial?
4. What methods of defense do echinoderms have to protect themselves from predators?
5. **Think Critically** Why would the ability to regenerate lost body parts be an important adaptation for sea stars, brittle stars, and other echinoderms?

Skill Builder Activities

6. **Forming Hypotheses** In your Science Journal, write a hypothesis about why echinoderms live on the ocean floor. **For more help, refer to the Science Skill Handbook.**

7. **Communicating** Choose an echinoderm that is discussed in this section and write about it in your Science Journal. Describe the following: *its appearance, how it gets food, where it lives,* and *other interesting facts.* **For more help, refer to the Science Skill Handbook.**

Activity

What do worms eat?

Earthworms are valuable because they improve the soil in which they live. There can be 50,000 earthworms living in one acre. Their tunnels increase air movement through the soil and improve water drainage. As they eat the decaying material in soil, their wastes can enrich the soil. Other than decaying material, what else do earthworms eat? Do they have favorite foods?

What You'll Investigate
What types of foods do earthworms eat?

Goals
- **Construct** five earthworm habitats.
- **Test** different foods to determine which ones earthworms eat.

Safety

WARNING: *Do not handle earthworms with dry hands. Do not eat any materials used in this activity.*

Materials
orange peels
apple peels
banana skin
kiwi fruit skin
watermelon rind
*skins of five different fruits
wide mouth jar (5)
potting soil
water
humus
*peat moss
earthworms
black construction paper (5 sheets)
masking tape
marker
rubber bands (5)
*Alternate materials

62 ◆ C CHAPTER 2 Mollusks, Worms, Arthropods, Echinoderms

Using Scientific Methods

Procedure

1. Copy the data table below in your Science Journal.
2. Pour equal amounts of soil into each of the jars. Do not pack the soil. Leave several centimeters of space at the top of each jar.
3. Sprinkle equal amounts of water into each jar to moisten the soil. Avoid pouring too much water into the jars.
4. Pour humus into each of your jars to a depth of 2 cm. The humus should be loose.
5. Add watermelon rinds to the first jar, orange peels to the second, apple peels to the third, kiwi fruit skins to the fourth, and a banana peel to the fifth jar. Each jar should have 2 cm of fruit skins on top of the layer of humus.
6. Add five earthworms to each jar.
7. Wrap a sheet of black construction paper around each jar and secure it with a rubber band.
8. Using the masking tape and marker, label each jar with the type of fruit it contains.
9. Place all of your jars in the same cool, dark place. Observe your jars every other day for a week and **record** your observations in your data table.

Fruit Wastes					
Date	Watermelon rind	Orange peels	Apple peels	Kiwi skins	Banana peels

Conclude and Apply

1. **Compare** the amount of fruit skins left in each jar.
2. **Infer** the type of food favored by earthworms.
3. **Infer** why some of the fruit skins were not eaten by the earthworms.
4. **Identify** a food source in each jar other than the fruit skins.
5. **Predict** what would happen in the jars over the next month if you continued the experiment.

Use the results of your experiment and information from your reading to help you write a recipe for an appetizing dinner that worms would enjoy. Based on the results of your experiment, add other fruit skins or foods to your menu you think worms would enjoy.

Science and Language Arts

"The Creature of My Mind"
by Ursula K. Le Guin

Respond to the Reading

1. How do you suppose the beetle injured itself?
2. What kind of insects does the author fear?
3. From the author's description, in what stage of development is the beetle?

When I stayed for a week in New Orleans… I had an apartment with a balcony… But when I first stepped out on it, the first thing I saw was a huge beetle. It lay on its back directly under the light fixture. I thought it was dead, then saw its legs twitch and twitch again. Big insects horrify me. As a child I feared moths and spiders, but adolescence cured me, as if those fears evaporated in the stew of hormones. But I never got enough hormones to make me easy with the large, hard-shelled insects: wood roaches, June bugs, mantises, cicadas. This beetle was a couple of inches long; its abdomen was ribbed, its legs long and jointed; it was dull reddish brown; it was dying. I felt a little sick seeing it lie there twitching, enough to keep me from sitting out on the balcony that first day…And if I had any courage or common sense, I kept telling myself, I'd… put it out of its misery. We don't know what a beetle may or may not suffer…

Legends and Oral Traditions

Understanding Literature

Legends and Oral Traditions In the passage you just read, the author uses her personal experience to consider her connection to other living things. A personal experience narrative is one in which the author tells a story using his or her own experience. In this piece, the author recounts a minor event in her life when she happens upon a dying beetle. The experience allows the author to pose some important questions about another species and to think about how beetles might feel when they die. How do you think the beetle is feeling?

Science Connection The author names several arthropod species in the passage, including isects and an arachnid. Beetles, June bugs, mantises, cicadas, and moths are all insects. The spider is an arachnid. Of the arthropods the author names, can you tell which ones go through a complete metamorphosis?

Linking Science and Writing

Writing a Personal Experience Narrative Write about a personal experience. Use the experience to think about an important question or topic in your life. For example, you might write about an accident you had. The accident might have made you consider the importance of good health.

Career Connection

Animal Geneticist

Edward B. Lewis was part of a genetic research team that won the 1995 Nobel Prize in Physiology or Medicine. They investigated how genes controlled the development of a fruit fly, *Drosophila*, through its various stages of development. Lewis investigated how genes of the fruit fly could control the development of specific regions into organs. Lewis also discovered that the genes were arranged on the chromosomes in the same order as the body parts that they controlled. For instance, the genes on one end of a complex strand controlled the development of the head, the genes in the middle controlled the development of the abdominal section, and the genes on the end controlled the development of the tail region. These genetic discoveries were a significant breakthrough because many of the same principles discovered in the fruit fly were found to apply in humans as well.

SCIENCE*Online* To learn more about careers in genetics, visit the Glencoe Science Web site at **science.glencoe.com**.

Chapter 2 Assessment

Checking Concepts

Choose the word or phrase that best answers the question.

1. What structure covers organs of mollusks?
   A) gills
   B) food
   C) mantle
   D) visceral mass

2. What structures do echinoderms use to move and to open shells of mollusks?
   A) mantle
   B) calcium plates
   C) spines
   D) tube feet

3. Which organism has a closed circulatory system?
   A) earthworm
   B) octopus
   C) slug
   D) snail

4. What evidence suggests that arthropods might have evolved from annelids?
   A) Arthropods and annelids have gills.
   B) Both groups have species that live in salt water.
   C) Segmentation is present in both groups.
   D) All segmented worms have setae.

5. Which of the following characteristics is typical of echinoderms?
   A) an endoskeleton
   B) a mantle
   C) a segmented body
   D) a water-vascular system

6. How do millipedes differ from centipedes?
   A) Millipedes are terrestrial and segmented.
   B) Millipedes eat plants.
   C) Millipedes have only one pair of legs on each segment.
   D) Millipedes have poison fangs.

7. Of the following organisms, which have two body regions and four pairs of legs?
   A) annelids
   B) arachnids
   C) insects
   D) mollusks

8. Which is an example of an annelid?
   A) earthworm
   B) octopus
   C) slug
   D) snail

9. Which group of animals is the largest?
   A) annelids
   B) arthropods
   C) echinoderms
   D) mollusks

10. Which sequence shows incomplete metamorphosis?
    A) egg—larvae—adult
    B) egg—nymph—adult
    C) larva—pupa—adult
    D) nymph—pupa—adult

Thinking Critically

11. Describe how the slug in the photo obtains food.

12. Compare the ability of clams, oysters, scallops, and squid to protect themselves.

13. Compare an earthworm gizzard to teeth in other animals.

14. What evidence suggests that mollusks and annelids share a common ancestor?

15. After molting but before the new exoskeleton hardens, an arthropod causes its body to swell by taking in extra water or air. How does this behavior help the arthropod?

Developing Skills

16. **Classifying** Place the following animals into arthropod groups: *spider, pill bug, crayfish, grasshopper, crab, silverfish, cricket, wasp, scorpion, shrimp, barnacle, tick,* and *butterfly.*

17. **Recognizing Cause and Effect** If all the earthworms were removed from a hectare of soil, what would happen to the soil? Why?

Chapter 2 Assessment

18. **Interpreting Scientific Illustrations** Using the illustrations in *Section 1*, infer why gastropods are sometimes called univalves.

19. **Researching Information** The suffixing *-ptera* means "wings." Research the meaning of the prefix of each insect group listed below and give an example of a member of each group.

   Diptera   Homoptera
   Orthoptera Hemiptera
   Coleoptera

20. **Comparing and Contrasting** Complete this Venn diagram to compare arthropods to annelids. What characteristics do they have in common? How are they different?

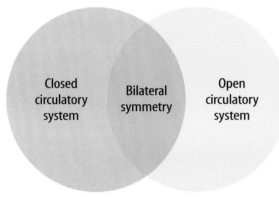

Performance Assessment

21. **Construct** Choose an arthropod that develops through complete metamorphosis and construct a three-dimensional model for each of the four stages. Share your construction with your class.

TECHNOLOGY

Go to the Glencoe Science Web site at **science.glencoe.com** or use the **Glencoe Science CD-ROM** for additional chapter assessment.

 **Test Practice**

Antonio researched mollusks and made the following circle graph, which separates mollusk species into seven living groups.

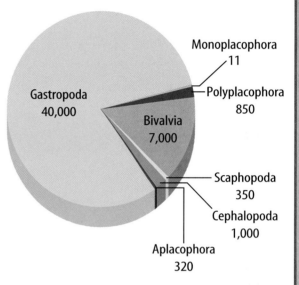

Study the circle graph and answer the following questions.

1. According to this information, approximately what percentage of mollusk species belongs to the group Gastropoda?
   - **A)** 40 percent
   - **B)** 60 percent
   - **C)** 80 percent
   - **D)** 95 percent

2. One of these groups was thought to be extinct until 1952, when marine biologists discovered some organisms living off the coast of Costa Rica. According to the graph, the group thought to be extinct is most likely the _____.
   - **F)** Bivalvia
   - **G)** Cephalopoda
   - **H)** Gastropoda
   - **J)** Monoplacophora

CHAPTER

3

Fish, Amphibians, and Reptiles

Do you know why frogs and salamanders live near ponds or streams? How do fish "breathe" underwater? What is the difference between alligators and crocodiles? In this chapter, you will find the answers to these questions. You also will read about the characteristics of animals known as chordates and vertebrates, and how fish, amphibians, and reptiles are classified, reproduce, and develop.

What do you think?

Science Journal Look at the picture below with a classmate. Discuss what you think this might be. Here's a hint: *It can be found in a pond during the spring season.* Write your answer or best guess in your Science Journal.

How much do you know about reptiles? For example, do snakes have eyelids? Why do snakes flick their tongues in and out? How can some snakes swallow animals that are larger than their own heads? Snakes don't have ears, so how do they hear? In this activity, you will discover the answer to one of these questions.

Model how a snake hears

1. Hold a tuning fork by the stem and tap it on a hard piece of rubber, such as the sole of a shoe.
2. Hold it next to your ear. What, if anything, do you hear?
3. Tap the tuning fork again. Press the base of the stem firmly against your chin. In your Science Journal, describe what happens.

Observe

Using the results from step 3, infer how a snake detects vibrations. In your Science Journal, predict how different animals can use vibrations to hear.

Before You Read

Making an Organizational Study Fold Make the following Foldable to help you organize your thoughts into clear categories about fish, amphibians, and reptiles.

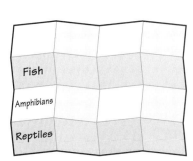

1. Place a sheet of paper in front of you so the short side is at the top. Fold the paper in half from the left side to the right side two times. Unfold all the folds.
2. Fold the paper in half from top to bottom. Then fold it in half again. Unfold all the folds.
3. Trace over all the fold lines. Label the rows *Fish, Amphibians,* and *Reptiles.*
4. As you read the chapter, write characteristics about each type of animal on your Foldable chart.

C ◆ 71

SECTION 1

Chordates and Vertebrates

As You Read

What You'll Learn
- **List** the characteristics of all chordates.
- **Identify** characteristics shared by vertebrates.
- **Differentiate** between ectotherms and endotherms.

Vocabulary
chordate
notochord
postanal tail
nerve cord
gill slit
endoskeleton
cartilage
vertebrae
ectotherm
endotherm

Why It's Important
Humans are vertebrates. Other vertebrates play important roles in your life because they provide food, companionship, and labor.

Chordate Characteristics

During a walk along the seashore at low tide, you often can see jellylike masses of animals clinging to rocks. Some of these animals may be sea squirts, as shown in **Figure 1,** which is one of the many types of animals known as chordates (KOR dayts). **Chordates** have four characteristics that are present at some stage of their development—a notochord (NOH tuh cord), postanal tail, nerve cord, and gill slits.

Notochord All chordates have an internal **notochord** that supports the animal and extends along the upper part of its body, as shown in **Figure 2.** The notochord is flexible but firm because it is made up of fluid-filled cells that are enclosed in a stiff covering. The notochord also extends into the **postanal tail**—a muscular structure at the end of the developing chordate. Some chordates, such as fish, amphibians, reptiles, birds, and mammals, develop backbones that partly or entirely replace the notochord. They are called vertebrates. In some chordates such as the sea squirt and other tunicates, and the lancelets, the notochord is kept into adulthood.

Reading Check *What happens to the notochord as a bat develops?*

Figure 1
Sea squirts get their name because when they're taken out of the ocean, they squirt water out of their body. *What do you have in common with a sea squirt?*

Nerve Cord Above the notochord and along the length of a developing chordate's body is a tubelike structure called the **nerve cord,** also shown in **Figure 2.** As most chordates develop, the front end of the nerve cord enlarges to form the brain and the remainder becomes the spinal cord. These two structures become the central nervous system that develops into complex systems for sensory and motor responses.

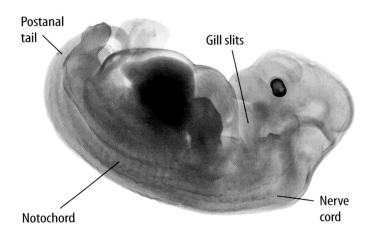

Figure 2
At some time during its development, a chordate has a notochord, postanal tail, nerve cord, and gill slits.

Gill Slits All developing chordates have **gill slits.** They are found in the region between the mouth and the digestive tube as pairs of openings to the outside. Many chordates have several pairs of gill slits. Ancient invertebrate chordates used their gill slits for filter feeding. This is still their purpose in some living chordates such as lancelets. In fish, the gill slits have developed into internal gills where oxygen and carbon dioxide are exchanged. In humans, gill slits are present only during embryonic development. However, one pair becomes the tubes that go from the ears to the throat.

Vertebrate Characteristics

Besides the characteristics common to all chordates, vertebrates have distinct characteristics. These traits set vertebrates apart from other chordates.

Structure All vertebrates have an internal framework called an **endoskeleton.** It is made up of bone and/or flexible tissue called **cartilage.** Your ears and the tip of your nose are made of cartilage. The endoskeleton provides a place for muscle attachment and supports and protects the organs. Part of the endoskeleton is a flexible, supportive column called the backbone, as shown in **Figure 3.** It is a stack of **vertebrae** alternating with cartilage. The backbone surrounds and protects the spinal nerve cord. Vertebrates also have a head with a skull that encloses and protects the brain.

Most of a vertebrate's internal organs are found in a central body cavity. A protective skin covers a vertebrate. Hair, feathers, scales, or horns sometimes grow from the skin.

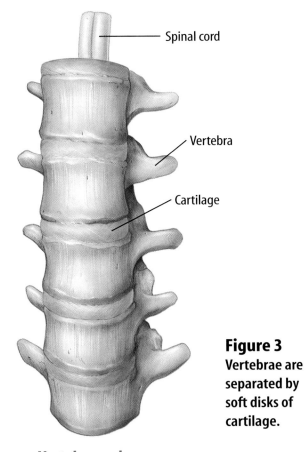

Vertebrae column

Figure 3
Vertebrae are separated by soft disks of cartilage.

SECTION 1 Chordates and Vertebrates C ◆ 73

Table 1 Types of Vertebrates

Group	Estimated Number of Species	Examples	
Jawless Fish	43	lamprey, hagfish	
Cartilaginous Fish	500 to 590	shark, ray, skate	
Bony Fish	20,000	salmon, bass, guppy, sea horse, lungfish	
Amphibians	4,000	frog, toad, salamander	
Reptiles	7,970	turtle, lizard, snake, crocodile, alligator	
Birds	8,700	stork, eagle, sparrow, turkey, duck, ostrich	
Mammals	4,600	human, whale, bat, mouse, lion, cow, otter	

Vertebrate Groups Seven main groups of vertebrates are found on Earth today, as shown in **Table 1.** Vertebrates are either ectotherms or endotherms. Fish, amphibians, and reptiles are ectotherms, also known as cold-blooded animals. An **ectotherm** has an internal body temperature that changes with the temperature of its surroundings. Birds and mammals are endotherms, which sometimes are called warm-blooded animals. An **endotherm** has a nearly constant internal body temperature.

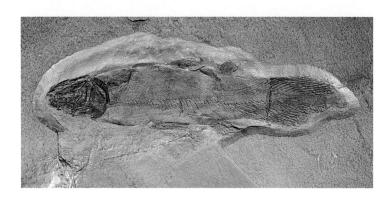

Figure 4
Placoderms were the first fish with jaws. These predatory fish were covered with heavy armor.

Vertebrate Origins Some vertebrate fossils, like the one in **Figure 4,** are of water-dwelling, armored animals that lived about 420 million years ago (mya). Lobe-finned fish appeared in the fossil record about 395 mya. The oldest known amphibian fossils date from about 370 mya. Reptile fossils have been found in deposits about 350 million years old. One well-known group of reptiles—the dinosaurs—first appeared about 230 mya.

In 1861, a fossil imprint of an animal with scales, jaws with teeth, claws on its front limbs, and feathers was found. The 150-million-year-old fossil was an ancestor of birds, *Archaeopteryx* (ar kee AHP tuh rihks).

Mammal-like reptiles appeared about 235 mya. However, true mammals appeared about 190 mya, and modern mammals originated about 38 million years ago.

Section Assessment

1. Name four characteristics that are shared by all chordates.
2. Explain the difference between a vertebra and a notocord.
3. What are characteristics of vertebrates that other chordates do not have?
4. What are some of the physical differences between ectotherm animals and endotherm animals?
5. **Think Critically** If the outside temperature decreases by 20°C, what will happen to a reptile's body temperature?

Skill Builder Activities

6. **Concept Mapping** Construct a concept map using these terms: *chordates, bony fish, amphibians, cartilaginous fish, reptiles, birds, mammals, lancelets, tunicates, invertebrate chordates, jawless fish,* and *vertebrates.* **For more help, refer to the** Science Skill Handbook.
7. **Communicating** Write a paragraph in your Science Journal comparing and contrasting the characteristics of chordates and vertebrates. **For more help, refer to the** Science Skill Handbook.

Activity

Endotherms and Ectotherms

Birds and mammals are endotherms. Fish, amphibians and reptiles are ectotherms. How can you determine whether an animal you have never seen before is an endotherm or an ectotherm? What tests might you conduct to find the answer?

What You'll Investigate
Is an animal that you construct an endotherm or an ectotherm?

Materials
fiberfill
*cotton balls
*old socks
*tissue
cloth
thermometer
*Alternate materials

Goals
- **Construct** an imaginary animal.
- **Determine** whether your animal is an endotherm or an ectotherm.

Safety Precautions

Procedure
1. Design an animal that has a thermometer inside. Construct the animal using cloth and some kind of stuffing material. Make sure that you will be able to remove and reinsert the thermometer.
2. Draw a picture of your animal and record data about its size and shape.

Animal Temperature		
Location	Beginning Time/ Temperature	Ending Time/ Temperature

3. Copy the data table in your Science Journal.
4. Place your animal in three locations that have different temperatures. Record the locations in the data table.

5. In each location, record the time and the temperature of your animal at the beginning and after 10 min.

Conclude and Apply
1. **Describe** your results. Did the animal's temperature vary depending upon the location?
2. Based on your results, is your animal an endotherm or an ectotherm? Explain.
3. **Compare** your results to those of others in your class. Were the results the same for animals of different sizes? Did the shape of the animal, such as one being flatter and another more cylindrical, matter?
4. Based on your results and what you read in the chapter, do you think your animal is most likely a bird, a mammal, a reptile, an amphibian, or a fish? Explain.

Communicating Your Data
Compare your conclusions with those of other students in your class. **For more help, refer to the** Science Skill Handbook.

SECTION 2 Fish

Fish Characteristics

Did you know that more differences appear among fish than among any other vertebrate group? In fact, there are more species of fish than species of other vertebrate groups. All fish are ectotherms. They are adapted for living in nearly every type of water environment on Earth—freshwater and salt water. Some fish, such as salmon, spend part of their life in freshwater and part of it in salt water. Fish are found at depths from shallow pools to deep oceans.

A streamlined shape, a muscular tail, and fins allow most fish to move rapidly through the water. **Fins** are fanlike structures attached to the endoskeleton. They are used for steering, balancing, and moving. Paired fins on the sides allow fish to move right or left and backward or forward. Fins on the top and bottom of the body give the fish stability. Most fish secrete a slimy mucus that also helps them move through the water.

Most fish have scales. **Scales** are hard, thin plates that cover the skin and protect the body, similar to shingles on the roof of a house. Most fish scales are made of bone. **Figure 5** illustrates how they can be tooth shaped, diamond shaped, cone shaped, or round. The shape of the scales can be used to help classify fish. The age of some species can be estimated by counting the annual growth rings of the scales.

As You Read

What You'll Learn
- **List** the characteristics of the three classes of fish.
- **Explain** how fish obtain food and oxygen and reproduce.
- **Describe** the importance and origin of fish.

Vocabulary
fin
scale

Why It's Important
Fish are an important food source for humans as well as many other animals.

Figure 5
Four types of fish scales are shown here.

A Sharks are covered with placoid scales such as these. Shark teeth are modified forms of these scales.

B Lobe-finned fish and gars are covered by ganoid scales. These scales don't overlap like other fish scales.

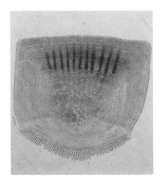

C Cycloid scales are thin and overlap, giving the fish flexibility. These scales grow as the fish grows.

D Ctenoid (TEN oyd) scales have a rough edge, which is thought to reduce drag as the fish swims through the water.

Figure 6
The sensory organs in the lateral line of a fish send messages to the fish's brain.

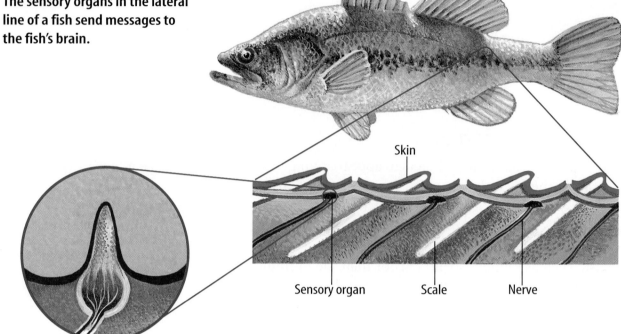

Figure 7
Even though a halibut's eyes are on one side of the fish, gills are on both sides.

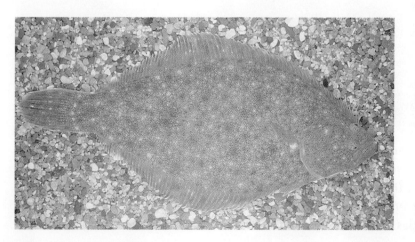

Body Systems All fish have highly developed sensory systems. Most fish have a lateral line system, as shown in **Figure 6.** A lateral line system is made up of a shallow, canal-like structure that extends along the length of the fish's body and is filled with sensory organs. The lateral line enables a fish to sense its environment and to detect movement. Some fish, such as sharks, also have a strong sense of smell. Sharks can detect blood in the water from several kilometers away.

Fish have a two-chambered heart in which oxygen-filled blood mixes with carbon dioxide-filled blood. A fish's blood isn't carrying as much oxygen as blood that is pumped through a three- or four-chambered heart.

Gas Exchange Most fish have organs called gills for the exchange of carbon dioxide and oxygen. Gills are located on both sides of the fish's head and are made up of feathery gill filaments that contain many tiny blood vessels. When a fish takes water into its mouth, the water passes over the gills, where oxygen from the water is exchanged with carbon dioxide in the blood. The water then passes out through slits on each side of the fish. Many fish, such as the halibut in **Figure 7,** are able to take in water while lying on the ocean floor.

Figure 8
Fish obtain food in different ways.

A A whale shark's mouth can open to 1.4 m wide.

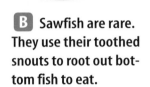

B Sawfish are rare. They use their toothed snouts to root out bottom fish to eat.

C Parrot fish use their hard beaks to bite off pieces of coral.

D Electric eels produce a powerful electric shock to stun their prey.

Feeding Adaptations Some of the adaptations that fish have for obtaining food are shown in **Figure 8**. Some of the largest sharks are filter feeders that take in small animals as they swim. The archerfish shoots down insects by spitting drops of water at them. Even though some fish have strong teeth, most do not chew their food. They use their teeth to capture their prey or to tear off chunks of food.

Reproduction Fish reproduce sexually. Reproduction is controlled by sex hormones. The production of sex hormones is dependent upon certain environmental factors such as temperature, length of daylight, and availability of food.

Female fish release large numbers of eggs into the water. Males then swim over the eggs and release sperm. This behavior is called spawning. The joining of the egg and sperm cells outside the female's body is called external fertilization. Certain species of sharks and rays have internal fertilization and lay fertilized eggs. Other fish, such as guppies, also have internal fertilization but do not lay eggs. The eggs develop and hatch inside the female's body. After they hatch, they leave her body.

Some species do not take care of their young. They release hundreds or even millions of eggs, which increases the chances that a few offspring will survive to become adults. Fish that care for their young lay fewer eggs. Some fish, including some catfish, hold their eggs and young in their mouths. Male sea horses keep the fertilized eggs in a pouch until they hatch.

Types of Fish

Fish vary in size, shape, color, living environments, and other factors. Despite their diversity, fish are grouped into only three categories—jawless fish, cartilaginous (kar tuh LA juh nuhs) fish, and bony fish.

Jawless Fish

Lampreys, along with the hagfish in **Figure 9,** are jawless fish. Jawless fish have round, toothed mouths and long, tubelike bodies covered with scaleless, slimy skin. Most lampreys are parasites. They attach to other fish with their suckerlike mouth. They then feed by removing blood and other body fluids from the host fish. Hagfish feed on dead or dying fish and other aquatic animals.

Jawless fish have flexible endoskeletons made of cartilage. Hagfish live only in salt water, but some species of lamprey live in salt water and other species live in freshwater.

Cartilaginous Fish

Sharks, skates, and rays are cartilaginous fish. These fish have skeletons made of cartilage like jawless fish. Unlike jawless fish, these fish have movable jaws that usually have well-developed teeth. Their bodies are covered with tiny scales that make their skins feel like fine sandpaper.

Sharks are top predators in many ocean food chains. They are efficient at finding and killing their food, which includes other fish, mammals, and some reptiles. Because of overfishing and the fact that shark reproduction is slow, shark populations are decreasing at an alarming rate.

✓ **Reading Check** *Why are shark populations decreasing?*

Health INTEGRATION

Many fish contain oil with omega-3 fatty acids, which seems to reverse the effects of too much cholesterol. A diet rich in fish that contain this oil might prevent the formation of fatty deposits in the arteries of humans. In your Science Journal, develop a menu for a meal that includes fish.

Figure 9
Hagfish have cartilaginous skeletons like sharks and rays. They feed on marine worms, mollusks, and crustaceans, in addition to dead and dying fish.

Figure 10
Bony fish come in many sizes, shapes, and colors. However, all bony fish have the same basic body structure.

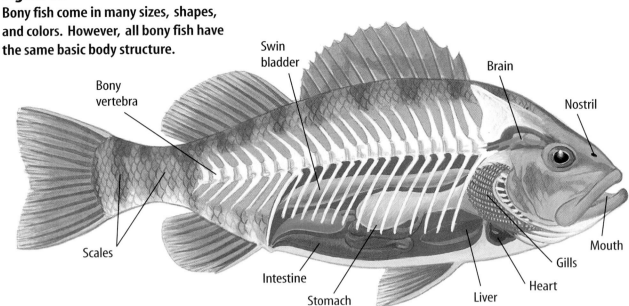

Bony Fish

About 95 percent of all species of fish are bony fish. They have skeletons made of bone. The body structure of a typical bony fish is shown in **Figure 10.** A bony flap covers and protects the gills. It closes as water moves into the mouth and over the gills. When it opens, water exits from the gills.

Swim Bladder An important adaptation in most bony fish is the swim bladder. It is an air sac that allows the fish to adjust its density in response to the density of the surrounding water. The density of matter is found by dividing its mass by its volume. If the density of the object is greater than that of the liquid it is in, the object will sink. If the density of the object is equal to the density of the liquid, the object will neither sink nor float to the surface. If the density of the object is less than the density of the liquid, the object will float on the liquid's surface.

The transfer of gases—mostly oxygen in deepwater fish and nitrogen in shallow-water fish—between the swim bladder and the blood causes the swim bladder to inflate and deflate. As the swim bladder fills with gases, the fish's density decreases and it rises in the water. When the swim bladder deflates, the fish's density increases and it sinks. Glands regulate the gas content in the swim bladder, enabling the fish to remain at a specific depth with little effort. Deepwater fish often have oil in their swim bladders rather than gases. Some bottom-dwelling fish and active fish that frequently change depth have no swim bladders.

Modeling How Fish Adjust to Different Depths

Procedure
1. Fill a **balloon** with air.
2. Place it in a **bowl of water.**
3. Fill another balloon partially with water, then blow air into it until it is the same size as the balloon filled only with air.
4. Place the second balloon in the bowl of water.

Analysis
1. What structure do these balloons model?
2. Compare where in the water (on the surface, or below the surface) two fish would be if they had swim bladders similar to the two balloons.

Lobe-finned Fish One of the three types of bony fish is the lobe-finned fish, as shown in **Figure 11**. Lobe-finned fish have fins that are lobelike and fleshy. These organisms were thought to have been extinct for more than 70 million years. But in 1938, some South African fishers caught a lobe-finned fish in a net. Several living lobe-finned fish have been studied since. Lobe-finned fish are important because scientists hypothesize that fish similar to these were the ancestors of the first land vertebrates—the amphibians.

Figure 11
Coelacanths (SEE luh kanthz) have been found living in the Indian Ocean north of Madagascar.

Math Skills Activity

Calculating Density

Example Problem

A freshwater fish has a mass of 645 g and a volume of 700 cm^3. What is the fish's density, and will it sink or float in freshwater?

Solution

1 *This is what you know:* density of freshwater = 1g/cm^3
mass of fish = 645 g
volume of fish = 700 cm^3

2 *This is what you need to find:* density of fish

3 *This is the equation you need to use:* $\dfrac{\text{mass of object (g)}}{\text{volume of object (cm}^3\text{)}}$ = density of object (g/cm^3)

4 *Substitute the known values:* $\dfrac{645 \text{ g}}{700 \text{ cm}^3}$ = 0.92 g/cm^3

The fish will float in freshwater. Its density is less than that of freshwater.

Practice Problem

Calculate the density of a saltwater fish that has a mass of 215 g and a volume of 180 cm^3. Will this fish float or sink in salt water? The density of ocean salt water is about 1.025 g/cm^3.

For more help, refer to the **Math Skill Handbook.**

Figure 12
Australian lungfish are one of the six species of lungfish.

Lungfish A lungfish, as shown in **Figure 12,** has one lung and gills. This adaptation enables them to live in shallow waters that have little oxygen. The lung enables the lungfish to breathe air when the water evaporates. Drought conditions stimulate lungfish to burrow into the mud and cover themselves with mucus until water returns. Lungfish have been found along the coasts of South America and Australia.

Ray-Finned Fish Most bony fish have fins made of long, thin bones covered with skin. Ray-finned fish, like those in **Figure 13,** have a lot of variation in their body plans. Most predatory fish have long, flexible bodies, which enable them to pursue prey quickly. Many bottom fish have flattened bodies and mouths adapted for eating off the bottom. Fish with unusual shapes, like the sea horse and anglerfish, also can be found. Yellow perch, tuna, salmon, swordfish, and eels are ray-finned fish.

Figure 13
Bony fish have a diversity of body plans.

A Most bony fish are ray-finned fish, like this rainbow trout.

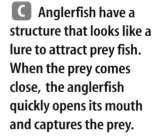

B Sea horses use their tails to anchor themselves to sea grass. This prevents the ocean currents from washing them away.

C Anglerfish have a structure that looks like a lure to attract prey fish. When the prey comes close, the anglerfish quickly opens its mouth and captures the prey.

Importance and Origin of Fish

Fish play a part in your life in many ways. They provide food for many animals, including humans. Fish farming and commercial fishing also are important to the U.S economy. Fishing is a method of obtaining food as well as a form of recreation enjoyed by many people. Many fish eat large amounts of insect larvae, such as mosquitoes, which keeps insect populations in check. Some, such as grass carp, are used to keep the plant growth from clogging waterways. Captive fish are kept in aquariums for humans to admire their bright colors and exotic forms.

Reading Check *How are fish helpful to humans?*

Figure 14
Lancelets are small, eel-like animals. They spend most of their time buried in the sand and mud at the bottom of the ocean.

Most scientists agree that fish evolved from small, soft-bodied, filter-feeding organisms similar to present-day lancelets, shown in **Figure 14.** The earliest fossils of fish are those of jawless fish that lived about 450 million years ago. Fossils of these early fish usually are found where ancient streams emptied into the sea. This makes it difficult to tell whether these fish ancestors evolved in freshwater or in salt water.

A group of fish—the placoderms—appeared about 420 million years ago. They flourished and dominated most aquatic ecosystems for about 60 million years, then vanished from the fossil record. Placoderms had cartilaginous skeletons and were the first fish to have jaws. They were probably the ancestors of modern cartilaginous fish and bony fish. Today, more than 400 families of bony fish exist.

Section Assessment

1. What are three characteristics of fish?
2. Name the three classes of fish and give an example of each.
3. How do jawless fish and cartilaginous fish take in food?
4. Describe the many ways that fish are important to humans.
5. **Think Critically** Female fish lay thousands of eggs. Why aren't lakes and oceans overcrowded with fish?

Skill Builder Activities

6. **Concept Mapping** Make an events-chain concept map to show what must take place for the fish to rise from the bottom to the surface of the lake. **For more help, refer to the Science Skill Handbook.**
7. **Using a Database** Use your computer to make a database of the characteristics of the three classes of fish. **For more help, refer to the Technology Skill Handbook.**

SECTION 3 Amphibians

Amphibian Characteristics

The word *amphibian* comes from the Greek word *amphibios,* which means "double life." They are well named, because amphibians spend part of their lives in water and part on land. Frogs, toads, and the salamander shown in **Figure 15** are examples of amphibians. What characteristics do these animals have that allow them to live on land and in water?

Amphibians are ectotherms. Their body temperature changes when the temperature of their surroundings changes. In cold weather, amphibians become inactive and bury themselves in mud or leaves until the temperature warms. This period of inactivity during cold weather is called **hibernation.** Amphibians that live in hot, dry environments become inactive and hide in the ground when temperatures become too hot. Inactivity during the hot, dry months is called **estivation.**

Reading Check *How are hibernation and estivation similar?*

Respiration Amphibians have moist skin that is smooth, thin, and without scales. They have many capillaries directly beneath the skin and in the lining of the mouth. This makes it possible for oxygen and carbon dioxide to be exchanged through the skin and the mouth lining. Amphibians also have small, simple, saclike lungs in the chest cavity for the exchange of oxygen and carbon dioxide. Some salamanders have no lungs and breathe only through their skin.

Circulation The three-chambered heart in amphibians is an important change from the circulatory system of fish. In the three-chambered heart, one chamber receives oxygen-filled blood from the lungs and skin, and another chamber receives carbon dioxide-filled blood from the body tissues. Blood moves from both of these chambers to the third chamber, which pumps oxygen-filled blood to body tissues and carbon dioxide-filled blood back to the lungs. Limited mixing of these two bloods occurs.

As You Read

What You'll Learn
- **Describe** the adaptations amphibians have for living in water and living on land.
- **List** the kinds of amphibians and the characteristics of each.
- **Explain** how amphibians reproduce and develop.

Vocabulary
hibernation
estivation

Why It's Important
Because amphibians are sensitive to changes in the environment, they can be used as biological indicators.

Figure 15
Salamanders often are mistaken for lizards because of their shape. However, like all amphibians, they have a moist, scaleless skin that requires them to live in a damp habitat.

Figure 16
Red-eyed tree frogs are found in forests of Central and South America. They eat a variety of food, including insects and even other frogs.

Reproduction Even though amphibians are adapted for life on land, they depend on water for reproduction. Because their eggs do not have a protective, waterproof shell, they can dry out easily, so amphibians must have water to reproduce.

Amphibian eggs are fertilized externally by the male. As the eggs come out of the female's body, the male releases sperm over them. In most species the female lays eggs in a pond or other body of water. However, many species have developed special reproductive adaptations, enabling them to reproduce away from bodies of water. Red-eyed tree frogs, like the ones in **Figure 16,** lay eggs in a thick gelatin on the underside of leaves that hang over water. After the tadpoles hatch, they fall into the water below, where they continue developing. The Sonoran Desert toad waits for small puddles to form in the desert during the rainy season. It takes tadpoles only two to 12 days to hatch in these temporary puddles.

Development Most amphibians go through a developmental process called metamorphosis (met uh MOR fuh sus). Fertilized eggs hatch into tadpoles, the stage that lives in water. Tadpoles have fins, gills, and a two-chambered heart similar to fish. As tadpoles grow into frogs and toads, they develop legs, lungs, and a three-chambered heart. **Figure 17** shows this life cycle.

The tadpole of some amphibian species, such as salamanders, are not much different from the adult stage. Young salamanders look like adult salamanders, but they have gills and usually a tail fin.

Figure 17
Amphibians go through metamorphosis as they develop.

B After hatching, most young amphibians, like these tadpoles, do not look like adult forms.

A Amphibian eggs are laid in a jellylike material to keep them moist.

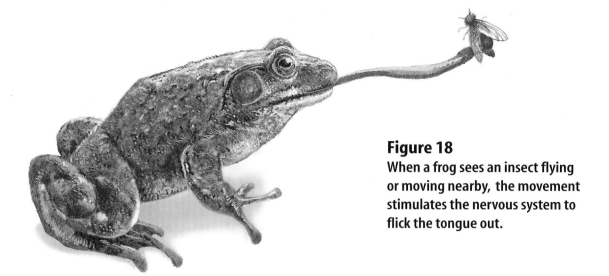

Figure 18
When a frog sees an insect flying or moving nearby, the movement stimulates the nervous system to flick the tongue out.

Frogs and Toads

Adult frogs and toads have short, broad bodies with four legs but no neck or tail. The strong hind legs are used for swimming and jumping. Bulging eyes and nostrils on top of the head let frogs and toads see and breathe while the rest of their body is submerged in water. On spring nights, they make their presence known with loud, distinctive croaking sounds. On each side of the head, just behind the eyes, are round tympanic membranes. These membranes vibrate somewhat like an eardrum in response to sounds and are used by frogs and toads to hear.

Frog and toad tongues are attached at the front of their mouths, as shown in **Figure 18.** When they see prey, their tongue flips out. When the tongue contacts the prey, the prey gets stuck in the sticky saliva on the tongue. Then the tongue flips back into the mouth. Toads and frogs eat a variety of insects, worms, and spiders, and one tropical species eats berries.

Research Visit the Glencoe Science Web site at **science.glencoe.com** for more information about amphibians as biological indicators. Communicate to your class what you learn.

C Amphibians go through metamorphosis, which means they change form from larval stage to adult.

D Most adult amphibians are able to move about and live on land.

SECTION 3 Amphibians

Mini LAB

Describing Frog Adaptations

Procedure

1. Carefully observe a **frog** in a **jar.** Notice the position of its legs as it sits. Record all of your observations in your **Science Journal.**
2. Observe its mouth, eyes, nostrils, and ears.
3. Observe the color of its back and belly.
4. Return the frog to your teacher.

Analysis

1. Describe the adaptations the frog has for living in water.
2. What adaptations does it have for living on land?

Figure 19
A Poison dart frogs are brightly colored to show potential predators that they are poisonous.
B Toxins from poison dart frogs have been used in hunting for centuries.

Salamanders

Most species of salamanders and newts live in North America. These amphibians often are mistaken for lizards because of their long, slender bodies. The short legs of salamanders and newts appear to stick straight out from the sides of their bodies.

Land-living species of salamanders and newts usually are found near water. These amphibians hide under leaf litter and rocks during the day to avoid the drying heat of the Sun. At night, they use their well-developed senses of smell and vision to find and feed on worms, crustaceans, and insects.

Many species of salamanders breed on land, where fertilization is internal. Aquatic species of salamanders and newts release and fertilize their eggs in the water.

Importance of Amphibians

Most adult amphibians are insect predators and are helpful in keeping some insect populations in check. They also are a source of food for other animals, including other amphibians. Some humans consider frog legs a delicacy.

Poison arrow frogs, like the one in **Figure 19,** produce a toxic poison that can kill large animals. They also are known as poison dart frogs. The toxin is secreted through their skin and can affect muscles and nerves of animals that come in contact with it. Native people of the Emberá Chocó in Colombia, South America, cover blowgun darts that they use for hunting with the poison of one species of these frogs. Researchers are studying the action of these toxins to learn more about how the nervous system works. Researchers also are using amphibians in regeneration studies in hopes of developing new ways of treating humans who have lost limbs or were born without limbs.

88 ◆ **C** **CHAPTER 3** Fish, Amphibians, and Reptiles

Biological Indicators Because they live on land and reproduce in water, amphibians are affected directly by any chemical change in the environment, including pesticides and other poisons, soil erosion, and water pollution. Amphibians also absorb gases through their skin, making them susceptible to air pollutants. Amphibians like the one in **Figure 20** are considered to be biological indicators. Biological indicators are species whose overall health reflects the health of a particular ecosystem.

 *What is a biological indicator?*

Figure 20
Beginning in 1995, deformed frogs such as this were found in Minnesota. Concerned scientists hypothesize that an increase in the number of deformed frogs could be a warning of environmental problems for other organisms.

Origin of Amphibians The fossil record shows that ancestors of modern fish were the first vertebrates on Earth about 500 million years ago. For about 150 million years, they were the only vertebrates. Then as competition for food and space increased and the climate changed, some lobe-finned fish might have traveled across land searching for water as their ponds dried up. The lobe-finned fish had lungs and bony fins that could have supported their weight on land. Amphibians are thought to have evolved from these lobe-finned fish about 350 million years ago.

Because competition on land from other animals was minimal, evolution favored the development of amphibians. Insects, spiders, and other invertebrates were an abundant source of food on land. Land was almost free of predators, so amphibians were able to reproduce in large numbers, and many new species evolved. For 100 million years or more, amphibians were the dominant land animals.

Section Assessment

1. List the adaptations amphibians have for living in water and for living on land.
2. Name three kinds of amphibians.
3. How do tadpole and frog hearts differ?
4. Describe two different environments where amphibians lay eggs.
5. **Think Critically** Why do you suppose frogs and toads seem to appear suddenly after a rain?

Skill Builder Activities

6. **Concept Mapping** Make an events-chain concept map of frog metamorphosis. Describe each stage in your Science Journal. **For more help, refer to the** Science Skill Handbook.
7. **Communicating** In your Science Journal, explain why frogs and other amphibians must live in moist or wet environments. **For more help, refer to the** Science Skill Handbook.

SECTION 4 Reptiles

As You Read

What You'll Learn
- **List** the characteristics of reptiles.
- **Determine** how reptile adaptations enable them to live on land.
- **Explain** the importance of the amniotic egg.

Vocabulary
amniotic egg

Why It's Important
Reptiles provide information about how body systems work during extreme weather conditions.

Figure 21
Some species of skinks, like this northern blue-tongue skink, don't lay eggs. The young develop inside the female and are born alive.

Reptile Characteristics

What do snakes and turtles have in common with dinosaurs? They, along with lizards, skinks like the one in **Figure 21,** crocodiles, and alligators, are reptiles. Reptiles have a variety of body shapes, but they have many characteristics in common that enable them to live on land.

Reptiles are ectotherms with a thick, dry, waterproof skin. Their skin is covered with scales that help reduce water loss and protect them from injury. Even though reptiles are ectotherms, they are able to modify their internal body temperatures by their behavior. When the weather is cold, they bask in the Sun, which is a behavior that enables them to warm up. When the weather is warm and the Sun gets too hot, they move into the shade to cool down.

Reading Check *How are reptiles able to modify their body temperature?*

Some reptiles, such as turtles, crocodiles, and lizards, move on four legs. Claws are used to dig, climb, and run. Reptiles, such as snakes and some lizards, move without legs.

Body Systems Scales on reptiles prevent the exchange of oxygen and carbon dioxide through the skin. Reptiles breathe with lungs. Even turtles and sea snakes that live in water must come to the surface to breathe.

The circulatory system of reptiles is more highly developed than that of amphibians. Most reptiles have a three-chambered heart with a partial wall inside the main chamber. This means that less mixing of oxygen-filled blood and carbon dioxide-filled blood occurs than in amphibians. This type of circulatory system provides more oxygen to all parts of the body. Crocodiles have a four-chambered heart that completely separates the oxygen-filled blood and the carbon dioxide-filled blood and keeps them from mixing.

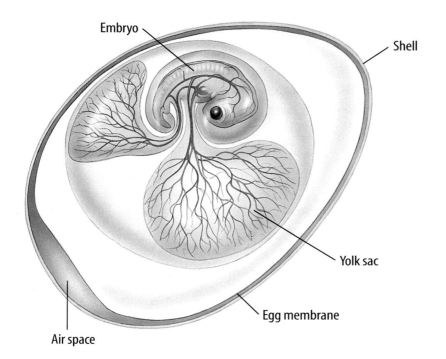

Figure 22
The development of amniotic eggs enabled reptiles to reproduce on land.

Amniotic Egg One of the most important adaptations of reptiles for living on land is the way they reproduce. Unlike the eggs of most fish and amphibians, eggs of reptiles are fertilized internally—inside the body of the female. After fertilization, the female lays eggs that are covered by tough, leathery shells. The shell prevents the eggs from drying out. This adaptation enables reptiles to lay their eggs on land.

The **amniotic egg** provides a complete environment for the embryo's development. **Figure 22** shows the structures in a reptilian egg. This type of egg contains membranes that protect and cushion the embryo, and help it get rid of wastes. It also contains a large food supply—the yolk—for the embryo. Minute holes in the shell, called pores, allow oxygen and carbon dioxide to be exchanged. By the time it hatches, a young reptile looks like a small adult.

 *What is the importance of an amniotic egg?*

Research Visit the Glencoe Science Web site at **science.glencoe.com** for recent news about the nesting sites of turtles. Communicate to your classmates what you learn.

Types of Modern Reptiles

Reptiles live on every continent except Antarctica and in all the oceans except those in the polar regions. They vary greatly in size, shape, and color. Reticulated pythons, 10 m in length, can swallow small deer whole. Some sea turtles have a mass of almost one metric ton and can swim faster than humans can run. Three-horned lizards have movable eye sockets and tongues as long as their bodies. The three living groups of reptiles are lizards and snakes, turtles, and crocodiles and alligators.

Lizards and Snakes Animals in the largest group of reptiles—the lizards and snakes like those shown in **Figure 23A** and **23B**—have a type of jaw not found in other reptiles, like the turtle in **Figure 23C**. The jaw has a special joint that unhitches and increases the size of their mouths. This enables them to swallow their prey whole. Lizards have movable eyelids, external ears, and legs with clawed toes on each foot. They feed on other reptiles, insects, spiders, worms, and mammals.

Snakes have developed ways of moving without legs. They have poor hearing and most have poor eyesight. Recall how you could feel the vibrations of the tuning fork in the Explore Activity. Snakes do not hear sound waves in the air. They "hear" vibrations in the ground that are picked up by the lower jawbone and conducted to the bones of the snake's inner ear. From there, the vibrations are transferred to the snake's brain, where the sounds are interpreted.

Snakes are meat eaters. Some snakes wrap around and constrict their prey. Others inject their prey with venom. Many snakes feed on rodents, and as a result help control rodent populations.

Most snakes lay eggs after they are fertilized internally. In some species, eggs develop and hatch inside the female's body then leave her body shortly thereafter.

Figure 23
Examples of reptiles are shown below.

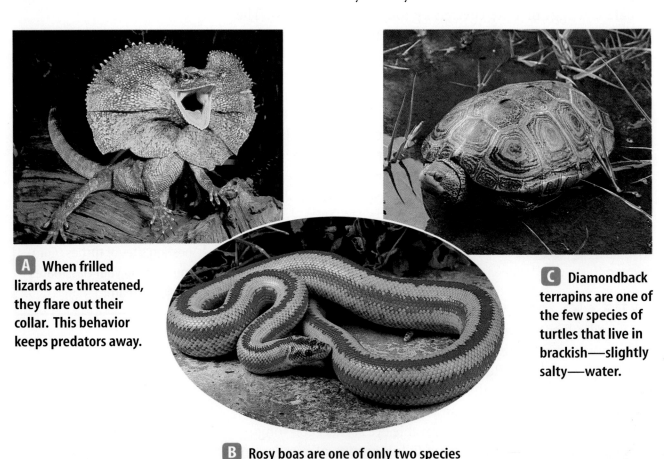

A When frilled lizards are threatened, they flare out their collar. This behavior keeps predators away.

C Diamondback terrapins are one of the few species of turtles that live in brackish—slightly salty—water.

B Rosy boas are one of only two species of boas found in the United States.

Turtles The only reptiles that have a two-part shell made of hard, bony plates are turtles. In some turtles, the shell is a tough leathery skin. The vertebrae and ribs are fused to the inside of the top part of the shell. The muscles are attached to the lower and upper part of the inside of the shell. Most turtles can withdraw their heads and legs into the shell for protection against predators.

✔ **Reading Check** *What is the purpose of a turtle's shell?*

Turtles have no teeth but they do have powerful jaws with a beaklike structure used to crush food. They feed on insects, worms, fish, and plants. Turtles live in water and on land. Those that live on land are called tortoises.

Like most reptiles, turtles provide little or no care for their young. Turtles dig out a nest, deposit their eggs, cover the nest, and leave. Turtles never see their own hatchlings. Young turtles, like those in **Figure 24**, emerge from the eggs fully formed and live on their own.

Figure 24
Most turtles are eaten shortly after they hatch. Only a few sea turtles actually make it into the ocean.

Crocodiles and Alligators Found in or near water in warm climates, crocodiles and alligators are similar in appearance. They are lizardlike in shape, and their backs have large, deep scales. Crocodiles and alligators can be distinguished from each other by the shape of their heads. Crocodiles like the one in **Figure 25** have a narrow head with a triangular-shaped snout. Alligators have a broad head with a rounded snout. Crocodiles are aggressive and can attack animals as large as cattle. Alligators are less aggressive than crocodiles. They feed on fish, turtles, and waterbirds. Crocodiles and alligators are among the world's largest living reptiles.

Crocodiles and alligators are some of the few reptiles that care for their young. The female guards the nest of eggs and when the eggs hatch, the male and female protect the young. A few crocodile females have been photographed opening their nests in response to noises made by hatchlings. After the young hatch, the females carry them in their huge mouths to the safety of the water. The female crocodile continues to keep watch over the young until they can protect themselves.

Figure 25
Indian gharials are one of the rarest crocodile species on Earth. Adults are well adapted for capturing fish.

NATIONAL GEOGRAPHIC VISUALIZING EXTINCT REPTILES

Figure 26

If you're like most people, the phrase "prehistoric reptiles" probably brings dinosaurs to mind. But not all ancient reptiles were dinosaurs. The first dinosaurs didn't appear until about 115 million years after the first reptiles. Paleontologists have unearthed the fossils of a variety of reptilian creatures that swam through the seas and waterways of ancient Earth. Several examples of these extinct aquatic reptiles are shown here.

▲ **MOSASAUR** (MOH zuh sawr) Marine-dwelling mosasaurs had snakelike bodies, large skulls, and long snouts. They also had jointed jawbones, an adaptation for grasping and swallowing large prey.

▲ **ICHTHYOSAUR** (IHK thee uh sawr) Ichthyosaurs resembled a cross between a dolphin and a shark, with large eyes, four paddlelike limbs, and a fishlike tail that moved from side to side. These extinct reptiles were fearsome predators with long jaws armed with numerous sharp teeth.

◀ **ELASMOSAURUS** (uh laz muh SAWR us) Predatory *Elasmosaurus* had a long neck— with as many as 76 vertebrae—topped by a small head.

▲ **CHAMPOSAUR** (CHAMP uh sawr) This ancient reptile looked something like a modern crocodile, with a long snout studded with razor-sharp teeth. Champosaurs lived in freshwater lakes and streams and preyed on fish and turtles.

▲ **PLESIOSAUR** (PLEE zee uh sawr) These marine reptiles had stout bodies, paddlelike limbs, and long necks. Plesiousaurs might have fed by swinging their heads from side to side through schools of fish.

The Importance of Reptiles

Reptiles are important predators in many environments. In farming areas, snakes eat rats and mice that destroy grains. Small lizards eat insects, and large lizards eat small animals that are considered pests.

Humans in many parts of the world eat reptiles and their eggs or foods that include reptiles, such as turtle soup. The number of reptile species is declining in areas where swamps and other lands are being developed for homes and recreation areas. Coastal nesting sites of sea turtles are being destroyed by development or are becoming unusable because of pollution. For years, many small turtles were collected in the wild and then sold as pets. People now understand that such practices disturb turtle populations. Today most species of turtles and their habitats are protected by law.

Earth Science INTEGRATION

Dinosaurs, reptiles that ruled Earth for 160 million years, died out about 65 million years ago. In your Science Journal, describe what changes in the environment could have caused the extinction of the dinosaurs.

Origin of Reptiles Reptiles first appeared in the fossil record about 345 million years ago. The earliest reptiles did not depend upon water for reproduction. As a result, they began to dominate the land about 200 million years ago. Some reptiles even returned to the water to live, although they continued to lay their eggs on land. Dinosaurs—descendants of the early reptiles—ruled Earth during this era, then died out about 65 million years ago. Some of today's reptiles, such as alligators and crocodiles, have changed little from their ancestors, some of which are illustrated in **Figure 26.**

Section 4 Assessment

1. What adaptations do reptiles have for living on land?
2. How do turtles differ from snakes, lizards, and alligators?
3. Why were early reptiles, including dinosaurs, so successful as a group?
4. Describe or draw the structure of an amniotic egg.
5. **Think Critically** Poisonous coral snakes and some harmless snakes have bright red, yellow, and black colors. How is this an advantage and a disadvantage to nonpoisonous snakes?

Skill Builder Activities

6. **Concept Mapping** Make a concept map showing the major characteristics of the types of reptiles. **For more help, refer to the** Science Skill Handbook.

7. **Solving One-Step Equations** Many dinosaurs were large. One large dinosaur, *Brachiosaurus,* was about 12 m tall and 22 m long. The largest land mammal now living is the elephant. The average size of an elephant is 3 m tall and 6 m long. How does the elephant compare in height to *Brachiosaurus*? **For more help, refer to the** Math Skill Handbook.

Activity: Design Your Own Experiment

Water Temperature and the Respiration Rate of Fish

What if last summer was hot with few storms? One day after many sunny, windless days, you noticed that a lot of dead fish were floating on the surface of your neighbor's pond. What might have caused these fish to die?

Recognize the Problem

How does water temperature affect the respiration rate of fish?

Form a Hypothesis

Fish obtain oxygen from the water. State a hypothesis about how water temperature affects the respiration rate of fish.

Goals

- **Design** and carry out an experiment to measure the effect of water temperature on the rate of respiration of fish.
- **Observe** the breathing rate of fish.

Possible Materials

goldfish
aquarium water
small fishnet
600-mL beakers
container of ice water
stirring rod
thermometer
aquarium

Safety Precautions

Protect your clothing. Use the fishnet to transfer fish into beakers.

Using Scientific Methods

Test Your Hypothesis

Plan

1. As a group, agree upon and write out the hypothesis statement. You might form a hypothesis that relates the amount of oxygen dissolved in water to water temperature and how this affects fish.

2. As a group, list the steps that you need to take to test your hypothesis. Be specific and describe exactly what you will do at each step. List your materials.

3. How will you measure the breathing rate of fish?

4. **Explain** how you will change the water temperature in the beakers. Fish respond better to a gradual change in temperature than an abrupt change. How will you measure the response of fish to changes in water temperature?

5. What data will your group collect? Prepare a data table in your Science Journal to record the data you collect. How many times will you run your experiment?

6. Read over your entire experiment to make sure the steps are in logical order. Identify any constants, variables, and controls.

Do

1. Make sure your teacher approves your set up and your plan before you start.

2. Carry out the experiment according to the approved plan.

3. While the experiment is going on, write down any observations that you make and complete the data table in your Science Journal.

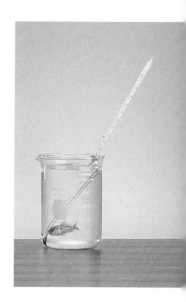

Analyze Your Data

1. **Compare** your results with the results of other groups in your class. Were the results similar?

2. What were you measuring when you counted mouth or gill cover openings?

3. **Describe** how a decrease in water temperature affects respiration rate and behavior of the fish.

4. **Explain** how your results could be used to determine the kind of environment in which a fish can live.

Draw Conclusions

1. Fish can live in water that is totally covered by ice. How is this possible?

2. What would happen to a fish if the water were to become very warm?

Construct a graph of your data on poster board and share your results with your classmates.

ACTIVITY C ◆ 97

TIME SCIENCE AND Society

SCIENCE ISSUES THAT AFFECT YOU!

The poison arrow frog

Ve

A poison found in some animals can save lives

Hiss, rattle… Run! Just the sound of a snake sends most people on a sprint to escape what could be a painful bite. Why? The bites could contain venom, a poisonous substance produced by certain species of animals. And venom can harm—or even kill—its victim. Some venomous creatures use their poison to stun, kill, and digest their prey, while others use it as a means of protection.

Venom is produced by a gland in the body. Some fish use their sharp, bony spines to inject venom. Venomous snakes, such as pit vipers, have fangs. Poison passes through these hollow teeth into a victim's body.

Puffer fish

The Gila monster, the largest lizard in the United States, has enlarged, grooved teeth in its lower jaw through which its venom travels. It is one of only two species of poisonous lizards. The poison arrow frog releases toxins through glands in the skin—predators immediately release these amphibians when they taste the poison. Puffer fish contain similar toxins but are only dangerous when eaten. Improper preparation of this fish can cause fatal poisoning. In some countries, chefs must be certified to prepare puffer fish.

The effects of venom on a victim can range from tingling, skin redness, and itching to paralysis and death. Some toxins cause vomiting or hallucinations. Hematoxic venom affects the blood, and can destroy cell tissue or rupture blood cells. Venoms containing neurotoxins are the most dangerous because they affect the nervous system. These toxins can paralyze the diaphragm, causing suffocation, or stop the heart from functioning.

How severely a victim is affected depends on more than the venom itself. A person's age, size, and general health play a role in the poison's effect. Also, bites and stings to fingers and toes are usually less serious than those to a person's head or body, because there's less of a chance that the venom will directly enter a blood vessel and the bloodstream.

Gila monster

Pit viper

Venom as Medicine

Doctors and scientists have discovered a shocking surprise within this sometimes deadly liquid. Oddly enough, the very same poison that harms and weakens people can heal, too. In fact, doctors use the deadliest venom—that of some pit viper species—to treat certain types of heart attacks. Cobra venom has been used to soothe the effects of cancer, and other snake venoms reduce the spasms of epilepsy and asthma.

Some venoms also contain substances that help blood clot. Hemophiliacs—people whose blood will not clot naturally—rely on the medical benefits that venom-based medicines supply. Venoms also are used in biological research. For instance, venoms that affect the nervous system help doctors and researchers learn more about how nerves function.

It's still smart to steer clear of the rattle or the stinger—but it's good to know that the venom in them might someday help as many as it can hurt.

CONNECTIONS Research Besides venom, what other defenses do animals use to protect themselves, or to subdue their prey? Explore how some animals native to your region use their built-in defenses.

SCIENCE *Online*
For more information, visit science.glencoe.com.

Chapter 3 Study Guide

Reviewing Main Ideas

Section 1 Chordates and Vertebrates

1. Chordates include lancelets, tunicates, and vertebrates. Chordates have a notochord, a nerve cord, gill slits, and a postanal tail.

2. All vertebrates have an endoskeleton that includes a backbone and a skull that protects the brain.

3. An endotherm is an animal that has a nearly constant internal body temperature. An ectotherm has a body temperature that changes with the temperature of its environment. *Is the animal in the photo an endotherm or an ectotherm?*

Section 2 Fish

1. Fish are vertebrates that have a streamlined body, fins, gills for gas exchange, and a highly developed sensory system.

2. Fish are divided into three groups—jawless fish, cartilaginous fish, and bony fish.

3. The bony fish have the greatest number of known fish species. Most bony fish have scales and swim bladders. *How does this fish use its swim bladder?*

Section 3 Amphibians

1. The first vertebrates to live on land were the amphibians.

2. Amphibians have adaptations that allow them to live on land and in the water. The adaptations include moist skin, mucous glands, and lungs. Most amphibians are dependent on water to reproduce. *Why is it necessary for this frog to live in a moist habitat?*

3. Most amphibians go through a metamorphosis from egg, to larva, to adult. During metamorphosis, legs develop, lungs replace gills, and the tail is lost.

Section 4 Reptiles

1. Reptiles are land animals with thick, dry, scaly skin. They lay amniotic eggs with leathery shells.

2. Turtles with tough shells, meat-eating crocodiles and alligators, and snakes and lizards make up the reptile groups.

3. Early reptiles were successful because of their adaptations to living on land. *What are some adaptations of this lizard?*

After You Read

To help you review the characteristics of fish, amphibians, and reptiles, use the Foldable you made at the beginning of the chapter.

Chapter 3 Study Guide

Visualizing Main Ideas

Complete the concept map below that describes chordates.

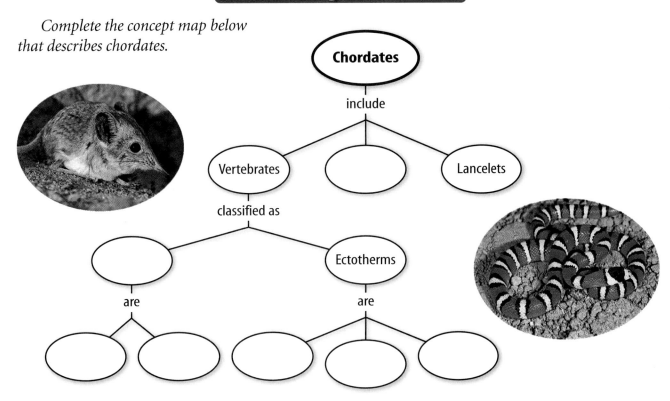

Vocabulary Review

Vocabulary Words

a. amniotic egg
b. cartilage
c. chordate
d. ectotherm
e. endoskeleton
f. endotherm
g. estivation
h. fin
i. gill slit
j. hibernation
k. nerve cord
l. notochord
m. postanal tail
n. scale
o. vertebrae

 **Study Tip**

Use tables and graphs to help you organize written material. For example, put the levels of biological organization in a table. Show what each level contains. Referring to a table can help you review concepts quickly.

Using Vocabulary

Replace the underlined word with the correct vocabulary word(s).

1. All chordates have a notochord, gill slits, postanal tail, and a <u>backbone</u>.
2. The inactivity of amphibians during hot, dry weather is <u>hibernation</u>.
3. All animals with a constant internal temperature are <u>vertebrates</u>.
4. Reptiles are <u>endotherms</u> with scaly skin.
5. Jawless fish have skeletons made of a tough, flexible tissue called <u>bone</u>.
6. Reptiles lay <u>eggs covered in gelatin</u>.
7. The structure that becomes the backbone in vertebrates is the <u>nerve cord</u>.

CHAPTER STUDY GUIDE C ◆ 101

Chapter 3 Assessment

Checking Concepts

Choose the word or phrase that best answers the question.

1. Which animals have fins, scales, and gills?
   A) amphibians   C) reptiles
   B) crocodiles   D) fish

2. Which is an example of a cartilaginous fish?
   A) hagfish   C) perch
   B) tuna   D) goldfish

3. What fish group has the greatest number of species?
   A) bony   C) cartilaginous
   B) jawless   D) amphibians

4. Which of the following is a fish with gills and lungs?
   A) shark   C) lungfish
   B) ray   D) perch

5. Biological indicators include which group of ectothermic vertebrates?
   A) amphibians   C) bony fish
   B) cartilaginous fish   D) reptiles

6. Which kinds of reptiles are included with lizards?
   A) snakes   C) turtles
   B) crocodiles   D) alligators

7. What term best describes eggs of reptiles?
   A) amniotic   C) jellylike
   B) brown   D) hard-shelled

8. Vertebrates that have lungs and moist skin belong to which group?
   A) amphibians   C) reptiles
   B) fish   D) lizards

9. How can crocodiles be distinguished from alligators?
   A) care of the young
   B) scales on the back
   C) shape of the head
   D) habitats in which they live

10. Which group has a notochord, nerve cord, postanal tail, and gill slits at some point in time?
    A) echinoderms   C) arthropods
    B) mollusks   D) chordates

Thinking Critically

11. Populations of frogs and toads are decreasing in some areas. What effects could this decrease have on other animal populations?

12. Why are some amphibians considered biological indicators?

13. In what ways are tunicates and lancelets similar to humans?

14. What physical features are common to all vertebrates?

15. Explain how the development of the amniotic egg led to the success of early reptiles.

Developing Skills

16. **Communicating** In your Science Journal, sequence the order in which these structures appeared in evolutionary history, then explain what type of organism had this adaptation and the advantage it provided: skin has mucous glands; skin has scales; dry, scaly skin.

17. **Comparing and Contrasting** Complete this chart that compares the features of some vertebrate groups.

Vertebrate Groups			
Feature	Fish	Amphibians	Reptiles
Circulation			
Respiration			
Reproduction			

Chapter 3 Assessment

18. Identifying and Manipulating Variables and Controls Design an experiment to find out the effect of water temperature on frog egg development.

19. Making and Using Graphs Make a circle graph of the species of fish in the table below. What percent of the graph is accounted for by bony fish?

Fish Species	
Kinds of Fish	Number of Species
Jawless	45
Cartilaginous	500
Bony	20,000

20. Classifying To what animal group does an animal with a two-chambered heart belong?

Performance Assessment

21. Conduct a Survey Many people are wary of reptiles. Write questions about reptiles to find out how people feel about these animals. Give the survey to your classmates, then graph the results and share them with your class.

22. Display Cut out pictures of fish from magazines and mount them on poster board. Letter the names of each fish on 3" × 5" cards. Have your classmates try to match the names of the fish with their pictures. To make this activity more challenging, use only the scientific names of each fish.

Technology

Go to the Glencoe Science Web site at **science.glencoe.com** or use the **Glencoe Science CD-ROM** for additional chapter assessment.

Test Practice

The following table shows the diversity of reptiles worldwide.

Worldwide Diversity of Reptiles	
Region	Reptile Species
Africa	1,320
Asia	2,080
Australia	850
Caribbean	500
Central America	1,060
Europe	200
Indian Ocean	360
Middle East	360
North America	340
Oceania	230
South America	1,560

Study the table shown above and answer the following questions.

1. The region that has the largest number of reptile species is _____ .
   A) North America
   B) Middle East
   C) Central America
   D) Asia

2. The number of classified reptile species worldwide is 7,984. The total of the table does not agree. Why?
   F) Scientists exaggerate the number of species they classify.
   G) Some areas share reptile species.
   H) Scientists keep poor records.
   J) Areas seldom share reptile species.

CHAPTER 4

Birds and Mammals

Why don't cats and dogs lay eggs? Why don't ostriches chew their food before they swallow it? Are the zebras and flamingos in the picture alike in any way? In this chapter, you will find the answers to these questions. Also presented in this chapter are the unique characteristics of birds and mammals. You will learn how birds and mammals are adapted to reproduce and live in many different environments.

What do you think?

Science Journal Look at the picture below with a classmate. Discuss what this might be or what is happening. Here's a hint: *It has found a protected place to eat and grow.* Write your answer or best guess in your Science Journal.

You may have observed a variety of animals in your neighborhood. Maybe you have watched birds at a bird feeder. Birds don't chew their food because they don't have teeth. Instead, many birds swallow small pebbles, bits of eggshells, and other hard materials that go into the gizzard—a muscular digestive organ. Inside the gizzard, they help grind up the seeds. Do the activity below as a model of a gizzard in action.

Model how a bird's gizzard works

1. Place some cracked corn, sunflower seeds, nuts or other seeds, and some gravel in an old sock.
2. Roll the sock on a hard surface and tightly squeeze it.
3. Describe the appearance of the seeds after rolling.

Observe
Describe in your Science Journal how a bird's gizzard helps it digest food.

Before You Read

Making a Compare and Contrast Study Fold As you study birds and mammals, use the following Foldable to help you compare and contrast their behavior.

1. Place a sheet of paper in front of you so that the long side is at the top. Fold in the left side and then the right side to divide the paper into thirds. Unfold the paper.
2. Fold the paper into thirds from top to bottom. Then fold it from the top to bottom again and unfold.
3. Trace over all the fold lines and label the table you created with *Birds* and *Mammals* at the top, as shown. In the column on the left list: *Vertebrate/Invertebrate, Diet, Movement, Body Symmetry,* and *Young,* as shown.
4. Before you read the chapter, write what you know about birds and mammals. As you read the chapter, add information to the table.

C ◆ 105

SECTION 1

Birds

As You Read

What You'll Learn
- **Identify** the characteristics of birds.
- **Identify** the adaptations birds have for flight.
- **Explain** how birds reproduce and develop.

Vocabulary
contour feather
down feather
endotherm
preening

Why It's Important
Most birds demonstrate structural and behavioral adaptations for flight.

Bird Characteristics

Birds are versatile animals. Geese have been observed flying at an altitude of 9,000 m, and penguins have been seen underwater at a depth of 543 m. An ostrich might weigh 155,000 g, while a hummingbird might weigh only 2 g. Some birds can live in the tropics and others can live in polar regions. Their diets vary and include meat, fish, insects, fruit, seeds, and nectar. Birds have feathers and scales and they lay eggs. Which of these characteristics is unique to birds?

Bird Eggs Birds lay amniotic (am nee AH tik) eggs with hard shells, as shown in **Figure 1A.** This type of egg provides a moist, protective environment for the developing embryo. The hard shell is made of calcium carbonate, the same chemical that makes up seashells, limestone, and marble. The egg is fertilized internally before the shell forms around it. The female bird lays one or more eggs usually in some type of nest, as shown in **Figure 1B.** A group of eggs is called a clutch. One or both parents may keep the eggs warm, or incubate them, until they hatch. The length of time for incubation varies from species to species. The young are cared for by one or both parents.

Figure 1
A nest's materials, shape, and location are different for different species of birds. For example, this robin's round nest is built of grasses and mud in a tree.

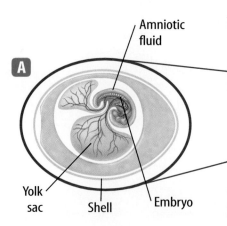

Figure 2
The hollow bones of birds are an adaptation for flight. *What advantages do the thin cross braces provide?*

Sternum The sternum has a structure called a keel, which is where flight muscles attach.

Tail A bird does not have a bony tail.

Leg bone

Hollow leg bone

Flight Adaptations People have always been fascinated by the ability of birds to fly. Flight in birds is made possible by their lightweight but strong skeleton, wings, feathers, strong flight muscles, and an efficient respiratory system. Well-developed senses, especially eyesight, and tremendous amounts of energy also are needed for flight.

Hollow Bones One adaptation that birds have for flight is an internal skeleton, as shown in **Figure 2.** Many bones of a bird often are joined together. This provides more strength and more stability for flight. Most bones of birds that fly are almost hollow. These bones have thin cross braces inside that strengthen the bones without adding much weight. The hollow spaces inside of the bones are filled with air.

 Reading Check *What features strengthen a bird's bones?*

A large sternum, or breastbone, supports the powerful chest muscles needed for flight. The last bones of the spine support the tail feathers, which play an important part in steering and balancing during flight and landing.

Astronomy INTEGRATION

Many theories have been proposed about how birds navigate at night. Some scientists hypothesize that star positions help night-flying birds find their way. Research the location of the North Star. In your Science Journal, infer how the North Star might help birds fly at night.

Mini LAB

Modeling Feather Preening

Procedure
1. Cut two 15-cm × 15-cm pieces of **cotton cloth.**
2. Apply **petroleum jelly** to one piece of the cloth.
3. Wet both pieces of the cloth with **water.**

Analysis
1. Compare the two pieces of cloth after they have been wet. In your **Science Journal,** describe what you observe. Wash your hands.
2. Infer why birds do not have to find shelter from the rain.

Feathers Birds are the only animals that have feathers. Their bodies are covered with two main types of feathers—contour feathers and down feathers. Strong, lightweight **contour feathers** give a bird its coloring and smooth shape. These are also the feathers that a bird uses when flying. The contour feathers on the wings and tail help the bird steer and keep it from spinning out of control.

Have you ever wondered how ducks can swim in a pond on a freezing cold day and keep warm? Soft, fluffy **down feathers** provide an insulating layer next to the skin of adult birds and cover the bodies of young birds. Birds are **endotherms,** meaning they maintain a constant body temperature. Feathers help birds maintain their body temperature, and grow in much the same way as your hair grows. Each feather grows from a microscopic pit in the skin called a follicle (FAHL ih kul). When a feather falls out, a new one grows in its place. As shown in **Figure 3,** the shaft of a feather has many branches called barbs. Each barb has many branches called barbules that give the feather strength.

Reading Check *Why are some young birds covered with down feathers?*

A bird has an oil gland located just above the base of its tail. Using its bill or beak, a bird rubs oil from the gland over its feathers in a process called **preening.** The oil conditions the feathers and helps make them water-repellent.

Figure 3
Down feathers help keep birds warm. Contour feathers are the feathers used for flight.

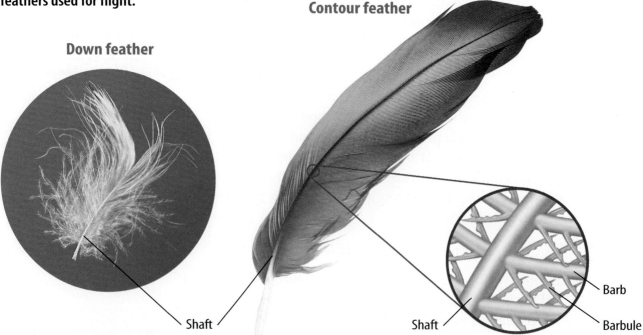

Figure 4
Wings provide an upward force called lift for birds and airplanes.

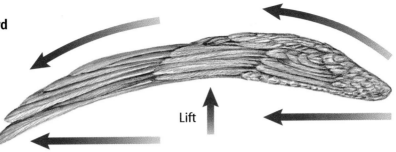

Lift

A Bald eagles are able to soar for long periods of time because their wings have a large surface area to provide lift.

B This glider gets lift from its wings the same way a bald eagle gets lift.

Wings Although not all birds fly, most wings are adapted for flight. Wings are attached to powerful chest muscles. By flapping its wings, a bird attains thrust to go forward and lift to stay in the air. Its wings move up and down, as well as back and forth.

The shape of a bird's wings helps it fly. The wings are curved on top and flat or slightly curved on the bottom. Humans copied this shape to make airplane wings, as shown in **Figure 4.** When a bird flies, air moves more slowly across the bottom than across the top of its wings. Slow-moving air has greater pressure than fast-moving air, resulting in an upward push called lift. The amount of lift depends on the total surface area of the wing, the speed at which air moves over the wing, and the angle of the wing to the moving air. Once birds with large wings, such as vultures, reach high altitudes they can soar and glide for a long time without having to beat their wings.

Wings also serve important functions for birds that don't fly. Penguins are birds that use their wings to swim underwater. Ostriches use their wings in courtship and to maintain their balance with their wings while running or walking.

Environmental Science INTEGRATION

Some birds have become pests in urban areas. Research to learn what birds are considered pests in urban areas, what affect they have on the urban environment, and what measures are taken to reduce the problems they create. Build a bulletin board of your results.

Research Visit the Glencoe Science Web site at **science.glencoe.com** for information about homing pigeons. Make a poster to illustrate how they are used.

Figure 5
A bird's blood is circulated quickly so enough oxygen-filled blood is carried to the bird's muscles.

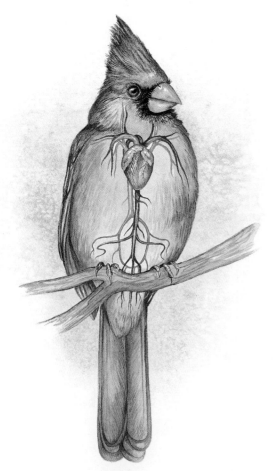

Body Systems

Whether they fly, swim, or run, most birds are extremely active. Their body systems are adapted for these activities.

Digestive System Because flying uses large amounts of energy, birds need large amounts of high energy foods, such as nuts, seeds, nectar, insects, and meat. Food is broken down quickly in the digestive system to supply this energy. In some birds, digestion can take less than an hour—for humans digestion can take more than a day.

From a bird's mouth, unchewed food passes into a digestive organ called the crop. The crop stores the food until it absorbs enough moisture to move on. The food enters the stomach where it is partially digested before it moves into the muscular gizzard. In the gizzard, food is ground and crushed by small stones and grit that the bird has swallowed. Digestion is completed in the intestine, and then the food's nutrients move into the bloodstream.

Respiratory System Body heat is generated when energy in food is combined with oxygen. A bird's respiratory system efficiently obtains oxygen, which is needed to power flight muscles and to convert food into energy. Birds have two lungs. Each lung is connected to balloonlike air sacs that reach into different parts of the body, including some of the bones. Most of the air inhaled by a bird passes into the air sacs behind the lungs. When a bird exhales, air with oxygen passes from these air sacs into the lungs. Air flows in only one direction through a bird's lungs. Unlike other vertebrates, birds receive air with oxygen when they inhale and when they exhale. This provides a constant supply of oxygen for the flight muscles.

Circulatory System A bird's circulatory system consists of a heart, arteries, capillaries, and veins, as shown in **Figure 5**. Their four-chambered heart is large compared to their body. On average, a sparrow's heart is 1.68 percent of its body weight. The average human heart is only 0.42 percent of the human's body weight. Oxygen-filled blood is kept separate from carbon dioxide-filled blood as both move through a bird's heart and blood vessels. A bird's heart beats rapidly—an active hummingbird's heart can beat more than 1,000 times per minute.

The Importance of Birds

Birds play important roles in nature. Some are sources of food and raw materials, and others are kept as pets. Some birds, like the owl in **Figure 6A,** help control pests, such as destructive rodents. Barn swallows and other birds help keep insect populations in check by eating them. Some birds, like the hummingbird in **Figure 6B,** are pollinators for many flowers. As they feed on the flower's nectar, pollen collects on their feathers and is deposited on the next flower they visit. Other birds eat fruits, then their seeds are dispersed in the birds' droppings. Seed-eating birds help control weeds. Birds can be considered pests when their populations grow too large. In cities where large numbers of birds roost, their droppings can damage buildings. Some droppings also can contain microorganisms that can cause diseases in humans.

Uses of Birds Humans have hunted birds for food and fancy feathers for centuries. Eventually, wild birds such as chickens and turkeys were domesticated and their meat and eggs became a valuable part of human diets. Feathers are used in mattresses and pillows because of their softness and ability to be fluffed over and over. Down feathers are good insulators. Even bird droppings, called guano (GWAH noh), are collected from sea bird colonies and used as fertilizer.

Parakeets, parrots, and canaries often are kept as pets because many sing or can be taught to imitate sounds and human voices. Most birds sold as pets are bred in captivity, but some wild birds still are collected illegally, which threatens many species.

Figure 6
In nature, some birds help control pests and others pollinate flowers. **A** Owls are birds of prey that hunt small animals at night. **B** As a hummingbird feeds on a flower's nectar, it may pollinate the flower.

NATIONAL GEOGRAPHIC VISUALIZING BIRDS

Figure 7

There are almost 9,000 living species of birds. Birds are subdivided into smaller groups based on characteristics such as beak size and shape, foot structure, and diet. Birds belonging to several groups are shown here.

INSECT-EATERS This nuthatch has a pointed beak that can pry up bark or bore into wood to find insects.

WATER BIRDS Wood ducks have webbed feet that propel them through the water.

BIRDS OF PREY This osprey has large claws that grasp and a sharp beak that tears flesh.

FLIGHTLESS BIRDS The ostrich evolved in places where there were once few mammal predators. Though they cannot fly, some flightless birds are fast runners.

WADING BIRDS The great blue heron's long legs allow it to walk in shallow water.

SEED-EATERS This cardinal's thick, strong beak can crack seeds.

Origin of Birds Birds, like those in **Figure 7,** have some characteristics of reptiles, including scales on their feet and legs. Scientists learn about the origins of most living things by studying their fossils. However, few fossils of birds have been found to study. Some scientists hypothesize that birds developed from reptiles millions of years ago.

Figure 8
A The first *Archaeopteryx* bones were found more than 100 years ago. *Archaeopteryx* is considered a link between reptiles and birds.
B *Protoavis* may be an ancestor of birds.

Archaeopteryx (ar kee AHP tuh rihks) is the oldest birdlike fossil—about 150 million years old. Although scientists do not think *Archaeopteryx* was a direct ancestor of modern birds, evidence shows that it had feathers and wings similar to today's birds. However, it had solid bones, teeth, a long bony tail, and clawed front toes, like some reptiles.

In 1991 in Texas, scientists discovered an older fossil that shows some characteristics of birds. *Protoavis* (proh toh AY vihs) had two characteristics of birds—hollow bones and a well-developed sternum with a keel. *Protoavis* lived about 225 million years ago. No fossil feathers were found with *Protoavis*. Scientists do not know if this animal was an ancestor of modern birds or a type of ground-living dinosaur. **Figure 8** shows an artist's idea of what *Archaeopteryx* and *Protoavis* may have looked like.

Section Assessment

1. List four characteristics shared by all birds.
2. Explain how a bird's feathers, air sacs, and skeleton are adaptations for flight.
3. What type of feather helps birds maintain their body temperature?
4. Make a network tree concept map that details the characteristics of birds. Use the following terms in your map: *birds, beaks, hollow bones, wings, eggs, adaptations for flight, feathers,* and *air sacs.*
5. **Think Critically** Hypothesize why most birds eat nuts, berries, insects, nectar, or meat, but not grass and leaves.

Skill Builder Activities

6. **Venn Diagram** Using information in this section, draw a Venn diagram in your Science Journal that compares and contrasts the characteristics of birds that fly and birds that do not fly. **For more help, refer to the** Science Skill Handbook.

7. **Communicating** Many expressions mention birds, such as "proud as a peacock" and "wise as an owl." In your Science Journal, make a list of several of these expressions and then decide which are accurate. **For more help, refer to the** Science Skill Handbook.

SECTION 2

Mammals

As You Read

What You'll Learn
- **Identify** the characteristics of mammals and explain how they have enabled mammals to adapt to different environments.
- **Distinguish** among monotremes, marsupials, and placentals.
- **Explain** why many species of mammals are becoming threatened or endangered.

Vocabulary
mammal
mammary gland
omnivore
carnivore
herbivore
monotreme
marsupial
placental
gestation period
placenta
umbilical cord

Why It's Important
Mammals, including humans, have many characteristics in common.

Characteristics of Mammals

You probably can name dozens of mammals, but can you list a few of their common characteristics? **Mammals** are endothermic vertebrates that have hair and produce milk to feed their young, as shown in **Figure 9**. Like birds, mammals care for their young. Mammals can be found almost everywhere on Earth. Each mammal species is adapted to its unique way of life.

Skin and Glands Skin covers and protects the bodies of all mammals. A mammal's skin is an organ that produces hair and in some species, horns, claws, nails, or hooves. The skin also contains different kinds of glands. One type of gland found in all mammals is the mammary gland. Female mammals have **mammary glands** that produce milk for feeding their young. Oil glands produce oil that lubricates and conditions the hair and skin. Sweat glands in some species remove wastes and help keep them cool. Many mammals have scent glands that secrete substances that can mark their territory, attract a mate, or be a form of defense.

Figure 9
Mammals, such as this moose, care for their young after they are born. *How do mammals feed their young?*

Figure 10
Mammals have teeth that are shaped specifically for the food they eat.

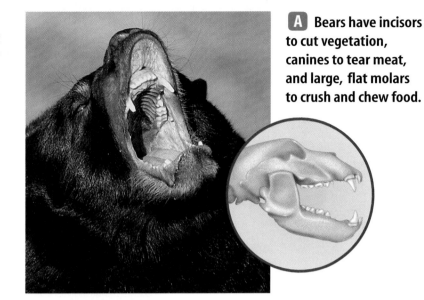

A Bears have incisors to cut vegetation, canines to tear meat, and large, flat molars to crush and chew food.

B A tiger easily can tear away the flesh of an animal because of large, sharp canine teeth and strong jaw muscles.

C A horse's back teeth, called molars, are large. *Infer how a horse chews.*

Teeth Notice that each mammal in **Figure 10** has different kinds of teeth. Almost all mammals have specialized teeth. Scientists can determine a mammal's diet by examining its teeth. Front teeth, called incisors, bite and cut. Sometimes the teeth next to the incisors, called canine teeth, are well developed to grip and tear. Premolars and molars at the back of the mouth shred, grind, and crush. Some animals, like the bear in **Figure 10A,** and humans, have all four kinds of teeth. They eat plants and other animals, so they are called **omnivores.** A **carnivore,** like the tiger in **Figure 10B,** has large canine teeth and eats only the flesh of other animals. **Herbivores,** such as the horse in **Figure 10C,** eat only plants. Their large premolars and molars grind the tough fibers in plants.

TRY AT HOME Mini LAB

Inferring How Blubber Insulates

Procedure
1. Fill a **self-sealing plastic bag** about ⅓ full with **vegetable shortening.**
2. Turn another self-sealing plastic bag inside out. Carefully place it inside the bag with the shortening so that you are able to seal one bag to the other. This is a blubber mitten.
3. Put your hand in the blubber mitten and place it in **ice water** for 5 s. Remove the blubber mitten when finished.
4. Put your bare hand in the same bowl of ice water for 5 s.

Analysis
1. Which hand seemed colder?
2. Infer how a layer of blubber provides protection against cold water.

Hair All mammals have hair on their bodies at some time during their lives. It may be thick fur that covers all or part of the animal's body. Fur traps air and helps to keep the animal warm. Whales have almost no hair. They rely on a thick layer of fat under their skin, called blubber, to keep them warm. Porcupine quills and hedgehog spines are modified hairs that offer protection from predators. Whiskers located near the mouth of many mammals, help them sense their environment.

Body Systems

The body systems of mammals are adapted to their activities and enable them to survive in many environments.

Mammals have four-chambered hearts that pump oxygen-filled blood directly throughout the body in blood vessels. Mammals have lungs made of millions of microscopic sacs. These sacs increase the lungs' surface area, allowing a greater exchange of carbon dioxide and oxygen.

A mammal's nervous system consists of a brain, spinal cord, and nerves. In mammals, the part of the brain involved in learning, problem solving, and remembering is larger than in other animals. Another large part of the mammal brain controls its muscle coordination.

The digestive systems of mammals vary according to the kinds of food they eat. Herbivores, like the one shown in **Figure 11,** have long digestive tracts compared to carnivores because plants take longer to digest than meat does.

Figure 11
Herbivores, like this elk, have four-chambered stomachs and long intestinal tracts that contain microorganisms, which help break down the plant material.

Reproduction and Caring for Young All mammals reproduce sexually. Most mammals give birth to live young after a period of development inside the female reproductive organ called the uterus. Many mammals are nearly helpless, and sometimes even blind, when they are born. They can't care for themselves for the first several days or even years. If you've seen newborn kittens or human babies, you know they just eat, sleep, grow, and develop. However, the young of some mammals, such as antelope, deer, and elephant are well developed at birth and are able to travel with their constantly moving parents. These young mammals usually can stand by the time they are a few minutes old. Marine mammals, such as dolphins and whales, as shown in **Figure 12,** can swim as soon as they are born.

 *Is a housecat or a deer more developed at birth?*

During the time that young mammals are dependent on their mother's milk, they learn many of the skills needed for survival. Defensive skills are learned while playing with other young of their own kind. In most mammal species, only females raise the young. However, males of some species, such as wolves and humans, help provide shelter, food, and protection for their young.

Figure 12
When a whale is born, the female whale must quickly push the newborn whale to the water's surface to breathe. Otherwise, the newborn whale will drown.

Problem-Solving Activity

Does a mammal's heart rate determine how long it will live?

Some mammals live long lives, but other mammals live for only a few years. Do you think that a mammal's life span might be related to how fast its heart beats? Use your ability to interpret a data table to answer this question.

Identifying the Problem
The table on the right lists the average heart rates and life spans of several different mammals. Heart rate is recorded as the number of heartbeats per minute, and life span is recorded as the average maximum years. When you examine the data, look for a relationship between the two variables.

Mammal Heart Rates and Life Spans		
Mammal	Heart Rate	Life Span
Mouse	400	2
Large dog	80	15
Bear	40	15-20
Elephant	25	75

Solving the Problem
1. Infer how heart rate and life span are related in mammals.
2. Humans have heart rates of about 70 beats per minute. Some humans may live for more than 100 years. Is this consistent with the data in the table? Explain.

SECTION 2 Mammals

Types of Mammals

Mammals are classified into three groups based on how their young develop. The three mammal groups are monotremes (MAHN uh treemz), marsupials (mar SEW pee ulz), and placentals (pluh SENT ulz).

Monotremes The duck-billed platypus, shown in **Figure 13,** is a well-known monotreme. Monotremes lay eggs with tough, leathery shells. The female incubates the eggs for about 10 days. After the young hatch, they nurse by licking the female's skin and hair where milk oozes from the mammary glands. Monotreme mammary glands do not have nipples.

Figure 13
A duck-billed platypus is a mammal, yet it lays eggs.
Why is it classified as a mammal?

Marsupials Many of the mammals that are classified as marsupials live in Australia, New Guinea, or South America. Only one type of marsupial, the opossum, lives in North America. **Marsupials** give birth to immature young that usually crawl into an external pouch on the female's abdomen. However, not all marsupials have pouches. Whether an immature marsupial is in a pouch or not, it instinctively crawls to a nipple. It stays attached to the nipple and feeds until it is developed. In pouched marsupials, the developed young return to the pouch for feeding and protection. Examples of marsupials are kangaroos and opossums, as shown in **Figure 14,** wallabies, koalas, bandicoots, and Tasmanian devils.

Figure 14
A Opossums are the only marsupials found in North America. **B** A joey, or young kangaroo, returns to its mother's pouch when danger is near.

Placentals In **placentals**, embryos completely develop inside the female's uterus. The time during which the embryo develops in the uterus is called the **gestation period.** Gestation periods range from 16 days in hamsters to 650 days in elephants. Placentals are named for the **placenta,** an organ that develops from tissues of the embryo and tissues that line the inside of the uterus. The placenta absorbs oxygen and food from the mother's blood. An umbilical cord connects the embryo to the placenta, as shown in **Figure 15.** Several blood vessels make up the umbilical cord. Blood in the **umbilical cord** transports food and oxygen from the placenta to the embryo and removes waste products from the embryo. The mother's blood doesn't mix with the embryo's blood. Examples of placentals are shown in **Table 1** on the following two pages.

Figure 15
An unborn mammal receives nutrients and oxygen through the umbilical cord.

 *Reading Check* **How does an embryo receive the things it needs to grow?**

Some placental groups include unusual animals such as the manatee shown in **Figure 16.** Dugongs and manatees are aquatic mammals. They have no back legs, and their front legs are modified into flippers. Another group includes small, rabbitlike animals called hyraxes that have hooves and molars for grinding vegetation. The aardvark is the only member of its group. Aardvarks have tubelike teeth and dig termites for food. Many Southeast Asian islands are home to members of a group which include gliding lemurs. Pangolins, another group of placentals, look like anteaters covered with scales.

Research Visit the Glencoe Science Web site at **science.glencoe.com** for recent news or magazine articles about manatees and their habitats. Communicate to your class what you learn.

Figure 16
A manatee swims slowly below the surface of the water.

SECTION 2 Mammals **C ◆ 119**

Table 1 Placentals

Order	Examples	Major Characteristics
Rodentia (roh DEN chuh)	beavers, mice, rats, squirrels	one pair of chisel-like front teeth adapted for gnawing; incisors growing throughout life; herbivores
Chiroptera (ki RAHP tuh ruh)	bats	front limbs adapted for flying; active at night; different species feed on fruit, insects, fish, or other bats
Insectivora (ihn sek TIHV uh ruh)	moles, shrews, hedgehogs	small, feed on insects, earthworms, and other small animals; most have long skulls and narrow snouts; high metabolic rate
Carnivora (kar NIHV uh ruh)	cats, dogs, bears, foxes, raccoons	long, sharp canine teeth for tearing flesh; most are predators, some are omnivores
Primates (PRI maytz)	apes, monkeys, humans	arms with grasping hands and opposable thumbs; eyes are forward facing; large brains; omnivores
Artiodactyla (ard ee oh DAHK tih luh)	deer, moose, pigs, camels, giraffes, cows	hooves with an even number of toes; most are herbivores with large, flat molars; complex stomachs and intestines

Order	Examples		Major Characteristics
Cetacea (sih TAY shuh)	whales, dolphins, porpoises		one or two blowholes on top of the head for breathing; forelimbs are modified into flippers; teeth or baleen
Lagomorpha (lag uh MOR fuh)	rabbits, hares, pikas		some with long hind legs adapted for jumping and running; one pair of large, upper incisors; one pair of small peglike incisors
Pinnipedia (pihn nih PEE dee uh)	sea lions, seals, walruses		marine carnivores; limbs modified for swimming
Edentata (ee dehn TAH tuh)	anteaters, sloths, armadillos		eat insects and other small animals; most toothless, or with tiny peglike teeth
Perissodactyla (puh ris oh DAHK tih luh)	horses, zebras, tapirs, rhinoceroses	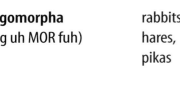	hooves with an odd number of toes; skeletons adapted for running; herbivores with large, grinding molars
Proboscidea (proh boh SIHD ee uh)	elephants		a long nose called a trunk; herbivores; upper incisor teeth grow to form tusks; thick, leathery skin

SECTION 2 Mammals C ◆ 121

Importance of Mammals

Mammals, like other organisms are important in maintaining a balance in the environment. Carnivores, such as lions, help control populations of other animals. Bats help pollinate flowers and control insects. Other mammals pick up plant seeds in their fur and distribute them. However mammals and other animals are in trouble today. As millions of wildlife habitats are destroyed for shopping centers, recreational areas, housing, and roads, many mammals are left without food, shelter, and space to survive. Because humans have the ability to reason, they have a responsibility to learn that their survival is related closely to the survival of all mammals. What can you do to protect the mammals in your community?

Origin of Mammals About 65 million years ago, dinosaurs and many other organisms became extinct. This opened up new habitats for mammals, and they began to branch out into many different species. Some of these species gave rise to modern mammals. Today, more than 4,000 species of mammals have evolved from animals similar to the one in **Figure 17,** which lived about 200 million years ago.

Figure 17
The Dvinia was an ancestor of mammals.

Section 2 Assessment

1. Describe five characteristics of all mammals and explain how they allow mammals to survive in different environments.
2. Differentiate among placentals, monotremes, and marsupials.
3. Name three ways in which mammals and birds are similar. Name three ways in which they are different.
4. Compare herbivores to omnivores.
5. **Think Critically** How have humans contributed to the decrease in many wildlife populations?

Skill Builder Activities

6. **Classifying** Classify the following animals into the three mammal groups: *whales, koalas, horses, elephants, opossums, kangaroos, rabbits, bats, bears, platypuses,* and *monkeys.* Compare and contrast their characteristics. **For more help, refer to the** Science Skill Handbook.

7. **Solving One-Step Equations** The tallest land mammal is the giraffe at 5.6 m. Calculate your height in meters and determine how many of you it would take to be as tall as the giraffe. **For more help, refer to the** Math Skill Handbook.

Activity

Mammal Footprints

Have you ever seen an animal footprint in the snow or soft soil? In this activity, you will observe pictures of mammal footprints and identify the mammal that made the footprint.

What You'll Investigate
How do mammal footprints differ?

Materials
diagram of footprints

Goals
- **Identify** mammal footprints.
- **Predict** where mammals live based on their footprints.

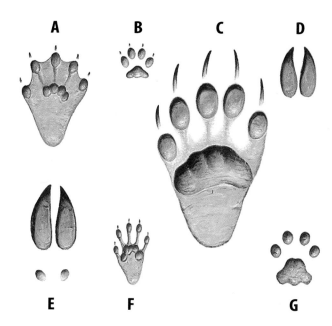

A B C D

E F G

Procedure

1. Copy the following data table in your Science Journal.

Identifying Mammal Footprints		
Animal	Letter of Footprint	Traits of Footprint
Bear		
Beaver		
Cougar		
Coyote		
Deer		
Moose		
Raccoon		

2. Compare and contrast the different mammal footprints in the above diagram.

3. Based on your observations, match each footprint to an animal listed in the first column of the data table.

4. **Write** your answers in the column labeled Letter of Footprint. Complete the data table.

Conclude and Apply

1. Which mammals have hoofed feet?
2. Which mammals have clawed toes?
3. Which mammals have webbed feet?
4. **Explain** how the different feet are adapted to the areas in which these different mammals live.
5. What are the differences between track **B** and track **E**? How does that help you identify the track?

Compare your conclusions with other students in your class. **For more help, refer to the** Science Skill Handbook.

ACTIVITY C ◆ 123

Activity
Use the Internet

Bird Counts

Birds can be found almost everywhere. You can see them in many different habitats—in a city park, an open field, along the riverbank, or at the shore. Many bird-watchers make their observations in the early morning when birds are most active. While bird-watching, care must be taken not to startle the birds with movement or noise.

It's simple to bird-watch. You can attract birds at home or at a school by filling a bird feeder with birdseed and placing it in sight of a window.

American Goldfinch

Cardinal

Black-capped Chickadee

Recognize the Problem
What is the most common bird in your neighborhood?

Form a Hypothesis
Think about the types of birds that you observe around your neighborhood. What types of food do they eat? Do all birds come to a bird feeder? Form a hypothesis about the type of bird that you think you will see most often at your bird feeder.

Goals
- **Research** how to build a bird feeder and attract birds to it.
- **Observe** the types of birds that visit your feeder.
- **Identify** the types of birds that you observe at your bird feeder.
- **Graph** your results and then communicate them to other students.

Data Source
SCIENCE *Online* Go to the Glencoe Science Web site at **science.glencoe.com** for more information about how to build a bird feeder, hints on bird-watching, and data from other students that do this activity.

Safety Precautions

124 ◆ C CHAPTER 4 Birds and Mammals

Using Scientific Methods

Test Your Hypothesis

Plan

1. **Research** general information about how to attract and identify birds. Determine where you will make your observations.
2. **Search** reference sources to find out how to build a bird feeder. Do all birds eat the same types of seeds?
3. What variables can you control in this activity? For what length of time will you make your observations? Do seasonal changes or weather conditions affect your observations?
4. What will you do to identify the birds that you do not know?

Do

1. Make sure your teacher approves your plan before you start.
2. **Record** your data in your Science Journal each time you make an observation of the birds at your bird feeder.

Birds at a feeder

Analyze Your Data

1. **Write** a description of where you placed your feeder and when you made your bird observations.
2. **Record** the total number of birds you observed each day.
3. **Record** the total number of each type of bird species you observed each day.
4. **Graph** your data using a line graph, a circle graph, or a bar graph.

Draw Conclusions

1. What type of bird was most common to your feeder?
2. Did all of your classmates' data agree with yours? Why or why not?
3. Review your classmates' data and determine if the location of bird observations affected the number of birds observed.
4. Did the time of day affect the number of birds observed? Explain.
5. Many birds eat great numbers of insects. Infer what humans might do to maintain a healthy environment for birds.

Communicating Your Data

SCIENCE Online Find this *Use the Internet* activity on the Glencoe Science Web site at **science.glencoe.com.** Post your data in the table provided. **Compare** your data to those of other students. Combine your data with those of other students and **plot** the combined data on a map to recognize patterns in bird populations.

ACTIVITY C ◆ 125

Science Stats

Eggciting Facts

Did you know...

...The ostrich lays the biggest egg of all birds now living. Its egg is 15 cm to 20 cm long and 10 cm to 15 cm wide. The volume of the ostrich egg is about equal to 24 chicken eggs. It can have a mass from approximately 1 kg to a little more than 2 kg. The shell of an ostrich egg is 1.5 mm thick and can support the weight of an adult human.

Ostrich egg

...The bird that lays the smallest egg is the hummingbird. Hummingbird eggs are typically 1.3 cm long and 0.8 cm wide. The smallest hummingbird egg on record was less than 1 cm long and weighed 0.36 g.

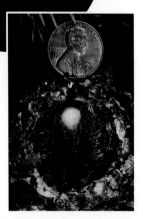

Hummingbird egg and nest

...The elephant bird, extinct within the last 1,000 years, laid an egg that was seven times larger than an ostrich egg. These eggs weighed about 12 kg. They were 30 cm long and could hold up to 8.5 L of liquid. It could hold the equivalent of 12,000 hummingbird eggs.

Elephant Bird/Vouron Patra *(Aepyornis maximus)*

Connecting To Math

...**The most valuable egg** in the world is called the Nicholas III Equestrian Egg. It is an antique egg made of enamel, silver, and gold. The egg is worth many dollars, and can be found in London, England.

...**May** is National Egg Month.

Eggs Eaten Per Person (1997)*

Country	Number
Brazil	111
Ireland	163
Japan	347
Russia	198
United States	238

*Average number

...**The shell of a chicken egg** can have 17,000 pores over its surface.

Do the Math

1. How many times longer is the largest ostrich egg than the smallest hummingbird egg?
2. How many elephant bird eggs would it take to equal to a dog weighing 48 kg?
3. Draw to scale an ostrich egg, a hummingbird egg, and an elephant bird egg.

Go Further

Visit the Glencoe Science Web site at **science.glencoe.com** and research the egg length of an American robin, a house sparrow, a bald eagle, and a Canada goose. Make a bar graph of this information.

SCIENCE STATS C ◆ 127

Chapter 4 Study Guide

Reviewing Main Ideas

Section 1 Birds

1. Birds are endothermic animals that are covered with feathers and lay eggs. *In the picture below, what type of feathers keep the goslings warm?*

2. Adaptations which enable most birds to fly include wings; feathers; a strong, lightweight skeleton; and efficient body systems. *Why can't these Emporer penguins fly?*

3. Birds lay eggs enclosed in hard shells. All birds' eggs are incubated until they hatch.

4. Birds help maintain a balance in nature by pollinating flowers and dispersing seeds, and by controlling animal pests and weeds.

5. Humans use birds and bird products for food, clothing, jewelry, bedding, fertilizer, and pets.

Section 2 Mammals

1. Mammals are endothermic animals with hair. *Why is this porcupine considered to be a mammal, even though it has quills?*

2. Female mammals have mammary glands that produce milk.

3. Mammals have teeth that are specialized for eating certain foods. Herbivores eat plants, carnivores eat meat, and omnivores eat plants and meat.

4. There are three groups of mammals. Monotremes lay eggs. Most marsupials have pouches for the development of their young. Placental offspring develop within a uterus and are nourished through a placenta. *The koala shown here has a pouch. To which group of mammals does it belong?*

5. Mammals, like other organisms, are important in maintaining balance in the environment. Their habitats are being destroyed.

After You Read

Use your Compare and Contrast Study Fold that you made at the beginning of this chapter to find similarities and differences between birds and mammals. What are some common characteristics of birds and mammals?

Chapter 4 Study Guide

Visualizing Main Ideas

Complete the following concept map on mammals.

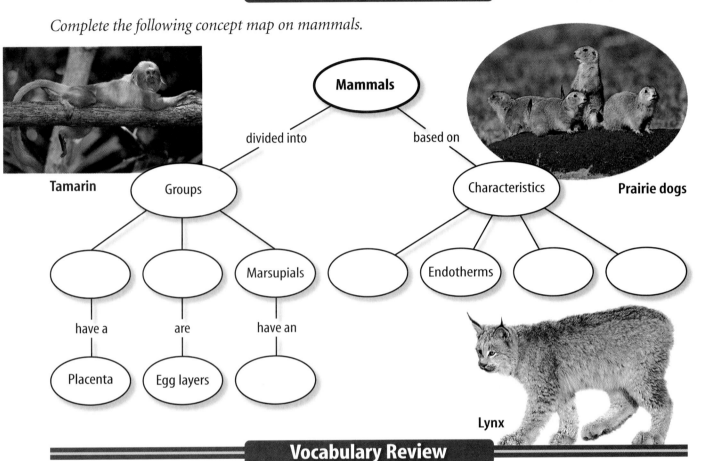

Vocabulary Review

Vocabulary Words

a. carnivore
b. contour feather
c. down feather
d. endotherm
e. gestation period
f. herbivore
g. mammal
h. mammary gland
i. marsupial
j. monotreme
k. omnivore
l. placenta
m. placentals
n. preening
o. umbilical cord

Using Vocabulary

Explain the difference between the vocabulary words in each of the following sets.

1. omnivore, carnivore, herbivore
2. contour feather, down feather
3. monotreme, marsupial
4. placenta, umbilical cord
5. endotherm, mammal
6. placental, monotreme
7. mammary gland, mammal
8. mammal, omnivore
9. endotherm, down feather
10. preening, down feather

 **Study Tip**

Ask what kinds of questions to expect on the test. Ask for practice tests so that you can become familiar with the test-taking materials.

CHAPTER STUDY GUIDE C ◆ 129

Chapter 4 Assessment

Checking Concepts

Choose the word or phrase that best answers the question.

1. Which of the following birds has feet adapted for moving on water?
   A) duck   C) owl
   B) oriole   D) rhea

2. Birds do NOT use their wings for which of the following activities?
   A) flying   C) balancing
   B) swimming   D) eating

3. Birds use which of the following organs to crush and grind their food?
   A) crop   C) gizzard
   B) stomach   D) small intestine

4. Which of the following are mammals that lay eggs?
   A) carnivores   C) monotremes
   B) marsupials   D) placentals

5. Which of the following mammals is classified as a marsupial?
   A) cat   C) kangaroo
   B) human   D) camel

6. Mammals with pouches are called what?
   A) marsupials   C) placentals
   B) monotremes   D) chiropterans

7. Which of the following have mammary glands without nipples?
   A) marsupials   C) monotremes
   B) placentals   D) omnivores

8. Teeth that are used for tearing food are called what?
   A) canines   C) molars
   B) incisors   D) premolars

9. Bird eggs do NOT have which of the following structures?
   A) hard shells   C) placentas
   B) yolks   D) membranes

10. Which of the following animals eat only plant materials?
    A) carnivores   C) omnivores
    B) herbivores   D) endotherms

Thinking Critically

11. Discuss the differences and similarities between bird reproduction and mammal reproduction.

12. You are a paleontologist studying fossils. One fossil appears to have hollow bones, a keeled breastbone, and a short, bony tail. How would you classify it?

13. Which type of bird would have lighter bones—a duck or an ostrich? Explain your answer.

14. What features of birds allow them to be fully adapted to life on land?

15. A mammal's teeth are similar in size and include all four types of teeth. What kind of mammal has teeth like this?

Developing Skills

16. **Classifying** Group the following mammals as herbivore, carnivore, or omnivore: *raccoon, mouse, rabbit, seal,* and *ape*.

17. **Making and Using Graphs** This table is a record of the approximate number of Canada geese that wintered at a midwestern wetland area over a five-year time period. Construct a line graph from these data.

| Record of Canada Geese ||
Year	Number of Geese
1996	550
1997	600
1998	575
1999	750
2000	825

Chapter 4 Assessment

18. Concept Mapping Complete this concept map about birds.

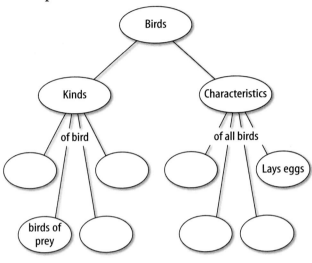

19. Classifying You discover three new species of placentals. The traits of each species are as follows.
Mammal 1 swims and eats meat.
Mammal 2 flies and eats fruit.
Mammal 3 runs on four legs and hunts.
Place each mammal into a correct order.

20. Comparing and Contrasting Describe the teeth of herbivores, carnivores, and omnivores. How are their types of teeth adapted to their diets?

Performance Assessment

21. Song with Lyrics Create a song about bird adaptations for flight by changing the words to a song that you know. Include in your song as many adaptations as possible.

TECHNOLOGY

Go to the Glencoe Science Web site at **science.glencoe.com** or use the **Glencoe Science CD-ROM** for additional chapter assessment.

 **Test Practice**

Students in Ms. Savir's science class placed five bird feeders at different locations around the school grounds. For five days, they recorded the number of each kind of bird that was attracted to the feeders. They plotted their results in the following line graph.

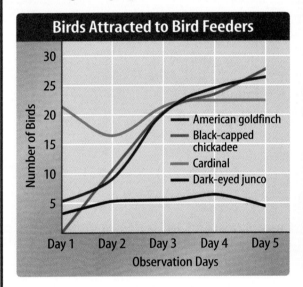

Study the graph and answer the following questions.

1. According to the graph, which bird did the students see the least?
   A) Black-capped chickadee
   B) Cardinal
   C) Dark-eyed junco
   D) American goldfinch

2. According to this information, which two birds were spotted on Day 3 the same number of times?
   F) American goldfinch and cardinal
   G) Black-capped chickadee and cardinal
   H) Dark-eyed junco and cardinal
   J) Black-capped chickadee and American goldfinch

CHAPTER ASSESSMENT   C ◆ 131

CHAPTER

5

Animal Behavior

Eye contact is made, dirt flies, and the silence is shattered. Massive horns clash as two bighorn sheep butt heads. Nearby, a spider spins a web to catch its food. Overhead, the honking of a V-shaped string of geese echoes through the valley. Do organisms learn these actions or do they occur automatically? In this chapter, you will examine the unique behaviors of animals. Also, you'll read about different types of behavior and learn about animal communication.

What do you think?
Science Journal Look at the picture below with a classmate. Discuss what you think this might be or what is happening. Here's a hint: *This instinctive reaction is triggered by their parent's arrival.* Write your answer or best guess in your Science Journal.

One way you communicate is by speaking. Other animals communicate without the use of sound. For example, a gull chick pecks at its parent's beak to get food. Try the activity below to see if you can communicate without speaking.

Observe how humans communicate without using sound

1. Form groups of students. Have one person choose an object and describe that object using gestures.
2. The other students observe and try to identify the object that is being described.

3. Each student in the group should choose an object and describe it without speaking while the others observe and identify the object.

Observe

In your Science Journal, describe how you and the other students were able to communicate without speaking to one another.

Before You Read

Making a Compare and Contrast Study Fold As you study behaviors, make the following Foldable to help find the similarities and differences between the behaviors of two animals.

1. Place a sheet of paper in front of you so the long side is at the top. Fold the paper in half from the left to the right side. Fold top to bottom but do not crease. Then unfold.
2. Label *Observed Behaviors of Animal 1* and *Observed Behaviors of Animal 2* across the front of the paper, as shown.
3. Through one thickness of paper, cut along the middle fold line to form two tabs, as shown.
4. Before you read the chapter, choose two animals to compare.
5. As you read the chapter, list the behaviors you learn about Animal 1 and Animal 2 under the tabs.

SECTION 1

Types of Behavior

As You Read

What You'll Learn
- **Identify** the differences between innate and learned behavior.
- **Explain** how reflexes and instincts help organisms survive.
- **Identify** examples of imprinting and conditioning.

Vocabulary
behavior
innate behavior
reflex
instinct
imprinting
conditioning
insight

Why It's Important
Innate behavior helps you survive on your own.

Behavior

When you come home from school, does your dog run to meet you? Your dog barks and wags its tail as you scratch behind its ears. Sitting at your feet, it watches every move you make. Why do dogs do these things? In nature, dogs are pack animals that generally follow a leader. They have been living with people for about 12,000 years. Dogs treat people as part of their own pack, as shown in **Figure 1.**

Animals are different from one another in their behavior. They are born with certain behaviors, and they learn others. **Behavior** is the way an organism interacts with other organisms and its environment. Anything in the environment that causes a reaction is called a stimulus. A stimulus can be external, such as a rival male entering another male's territory, or internal, such as hunger or thirst. You are the stimulus that causes your dog to bark and wag its tail. Your dog's reaction to you is a response.

Figure 1
Dogs are pack animals by nature. **A** This pack of wild dogs must work together to survive. **B** Domesticated dogs have accepted humans as their leader.

134 ◆ C CHAPTER 5 Animal Behavior

Innate Behavior

A behavior that an organism is born with is called an **innate behavior.** These types of behaviors are inherited. They don't have to be learned.

Innate behavior patterns occur the first time an animal responds to a particular internal or external stimulus. For birds like the swallows in **Figure 2A** and the hummingbird in **Figure 2B** building a nest is innate behavior. When it's time for the female weaverbird to lay eggs, the male weaverbird builds an elaborate nest, as shown in **Figure 2C.** Although a young male's first attempt may be messy, the nest is constructed correctly.

The behavior of animals that have short life spans is mostly innate behavior. Most insects do not learn from their parents. In many cases, the parents have died or moved on by the time the young hatch. Yet every insect reacts innately to its environment. A moth will fly toward a light, and a cockroach will run away from it. They don't learn this behavior. Innate behavior allows animals to respond instantly. This quick response often means the difference between life and death.

Reflexes The simplest innate behaviors are reflex actions. A **reflex** is an automatic response that does not involve a message from the brain. Sneezing, shivering, yawning, jerking your hand away from a hot surface, and blinking your eyes when something is thrown toward you are all reflex actions.

In humans a reflex message passes almost instantly from a sense organ along the nerve to the spinal cord and back to the muscles. The message does not go to the brain. You are aware of the reaction only after it has happened. Your body reacts on its own. A reflex is not the result of conscious thinking.

Figure 2
Bird nests come in different sizes and shapes. **A** Cliff swallows build nests out of mud. **B** Hummingbirds build delicate cup-shaped nests on branches of trees. **C** This male weaverbird is knotting the ends of leaves together to secure the nest.

Health INTEGRATION

A tap on a tendon in your knee causes your leg to stretch. This is known as the knee-jerk reflex. Abnormalities in this reflex tell doctors of a possible problem in the central nervous system. Research other types of reflexes and write a report about them in your Science Journal.

Figure 3
Spiders, like this orb weaver spider, know how to spin webs as soon as they are born.

Instincts An **instinct** is a complex pattern of innate behavior. Spinning a web like the one in **Figure 3** is complicated, yet spiders spin webs correctly on the first try. Unlike reflexes, instinctive behaviors can take weeks to complete. Instinctive behavior begins when the animal recognizes a stimulus and continues until all parts of the behavior have been performed.

✓ Reading Check   *What is the difference between a reflex and an instinct?*

Learned Behavior

All animals have innate and learned behaviors. Learned behavior develops during an animal's lifetime. Animals with more complex brains exhibit more behaviors that are the result of learning. However, the behavior of insects, spiders, and other arthropods is mostly instinctive behavior. Fish, reptiles, amphibians, birds, and mammals all learn. Learning is the result of experience or practice.

Learning is important for animals because it allows them to respond to changing situations. In changing environments, animals that have the ability to learn a new behavior are more likely to survive. This is especially important for animals with long life spans. The longer an animal lives, the more likely it is that the environment in which it lives will change.

Learning also can modify instincts. For example, grouse and quail chicks, shown in **Figure 4,** leave their nests the day they hatch. They can run and find food, but they can't fly. When something moves above them, they instantly crouch and keep perfectly still until the danger has passed. They will crouch without moving even if the falling object is only a leaf. Older birds have learned that leaves will not harm them, but they freeze when a hawk moves overhead.

Figure 4
As they grow older, these quail chicks will learn which organisms to avoid. *Why is it important for young quail to react the same toward all organisms?*

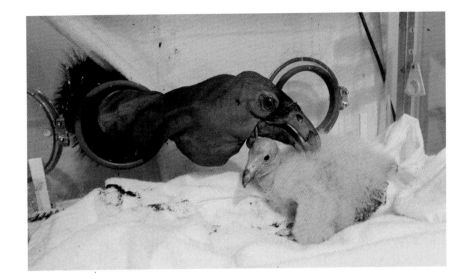

Figure 5
When feeding chicks in captivity, puppets of adult condors are used so the chicks don't associate humans with food.

Imprinting Learned behavior includes imprinting, trial and error, conditioning, and insight. Have you ever seen young ducks following their mother? This is an important behavior because the adult bird has had more experience in finding food, escaping predators, and getting along in the world. **Imprinting** occurs when an animal forms a social attachment, like the condor in **Figure 5,** to another organism within a specific time period after birth or hatching.

Konrad Lorenz, an Austrian naturalist, developed the concept of imprinting. Working with geese, he discovered that a gosling follows the first moving object it sees after hatching. The moving object, whatever it is, is imprinted as its parent. This behavior works well when the first moving object a gosling sees is an adult female goose. But goslings hatched in an incubator might see a human first and imprint on him or her. Animals that become imprinted toward animals of another species have difficulty recognizing members of their own species.

Research Visit the Glencoe Science Web site at **science.glencoe.com** for the latest information about raising condors to be released into the wild. Communicate to your class what you learn.

Trial and Error Can you remember when you learned to ride a bicycle? You probably fell many times before you learned how to balance on the bicycle. After a while you could ride without having to think about it. You have many skills that you have learned through trial and error such as feeding yourself and tying your shoes, as shown in **Figure 6.**

Behavior that is modified by experience is called trial-and-error learning. Many animals learn by trial and error. When baby chicks first try feeding themselves, they peck at many stones before they get any food. As a result of trial and error, they learn to peck only at food particles.

Figure 6
Were you able to tie your shoes on the first attempt? *What other things do you do every day that required learning?*

SECTION 1 Types of Behavior C ◆ 137

Mini LAB

Observing Conditioning

Procedure
1. Obtain several **photos of different foods and landscapes** from your teacher.
2. Show each picture to a classmate for 20 s.
3. Record how each photo made your partner feel.

Analysis
1. How did your partner feel after looking at the photos of food?
2. What effect did the landscape pictures have on your partner?
3. Infer how advertising might condition consumers to buy specific food products.

Figure 7
In Pavlov's experiment, a dog was conditioned to salivate when a bell was rung. It associated the bell with food.

Conditioning Do you have an aquarium in your school or home? If you put your hand above the tank, the fish probably will swim to the top of the tank expecting to be fed. They have learned that a hand shape above them means food. What would happen if you tapped on the glass right before you fed them? After a while the fish probably will swim to the top of the tank if you just tap on the glass. Because they are used to being fed after you tap on the glass, they associate the tap with food.

Animals often learn new behaviors by conditioning. In **conditioning,** behavior is modified so that a response to one stimulus becomes associated with a different stimulus. There are two types of conditioning. One type introduces a new stimulus before the usual stimulus. Russian scientist Ivan P. Pavlov performed experiments with this type of conditioning. He knew that the sight and smell of food made hungry dogs secrete saliva. Pavlov added another stimulus. He rang a bell before he fed the dogs. The dogs began to connect the sound of the bell with food. Then Pavlov rang the bell without giving the dogs food. They salivated when the bell was rung even though he did not show them food. The dogs, like the one in **Figure 7,** were conditioned to respond to the bell.

In the second type of conditioning, the new stimulus is given after the affected behavior. Getting an allowance for doing chores is an example of this type of conditioning. You do your chores because you want to receive your allowance. You have been conditioned to perform an activity that you may not have done if you had not been offered a reward.

Reading Check *How does conditioning modify behavior?*

Insight How does learned behavior help an animal deal with a new situation? Suppose you have a new math problem to solve. Do you begin by acting as though you've never seen it before, or do you use what you have learned previously in math to solve the problem? If you use what you have learned, then you have used a kind of learned behavior called insight. **Insight** is a form of reasoning that allows animals to use past experiences to solve new problems. In experiments with chimpanzees, as shown in **Figure 8,** bananas were placed out of the chimpanzees' reach. Instead of giving up, they piled up boxes found in the room, climbed them, and reached the bananas. At some time in their lives, the chimpanzees must have solved a similar problem. The chimpanzees demonstrated insight during the experiment. Much of adult human learning is based on insight. When you were a baby, you learned by trial and error. As you grow older, you will rely more on insight.

Figure 8
This illustration shows how chimpanzees may use insight to solve problems.

Section 1 Assessment

1. How is innate behavior different from learned behavior?
2. Compare a reflex with an instinct.
3. What is the difference between an internal and external stimulus?
4. Compare imprinting and conditioning.
5. **Think Critically** Use what you know about conditioning to explain how the term *mouthwatering food* might have come about.

Skill Builder Activities

6. **Researching Information** How are dogs trained to sniff out certain substances? **For more help, refer to the** Science Skill Handbook.
7. **Using an Electronic Spreadsheet** Make a spreadsheet of the behaviors in this section. Sort the behaviors according to whether they are innate or learned behaviors. Then identify the type of innate or learned behavior. **For more help, refer to the** Technology Skill Handbook.

SECTION 2

Behavioral Interactions

As You Read

What You'll Learn
- **Explain** why behavioral adaptations are important.
- **Describe** how courtship behavior increases reproductive success.
- **Explain** the importance of social behavior and cyclic behavior.

Vocabulary
social behavior
society
aggression
courtship behavior
pheromone
cyclic behavior
hibernation
migration

Why It's Important
Organisms must be able to communicate with each other to survive.

Instinctive Behavior Patterns

Complex interactions of innate behaviors between organisms result in many types of animal behavior. For example, courtship and mating within most animal groups are instinctive ritual behaviors that help animals recognize possible mates. Animals also protect themselves and their food sources by defending their territories. Instinctive behavior, just like natural hair color, is inherited.

Social Behavior

Animals often live in groups. One reason, shown in **Figure 9,** is that large numbers provide safety. A lion is less likely to attack a herd of zebras than a lone zebra. Sometimes animals in large groups help keep each other warm. Also, migrating animal groups are less likely to get lost than animals that travel alone.

Interactions among organisms of the same species are examples of **social behavior.** Social behaviors include courtship and mating, caring for the young, claiming territories, protecting each other, and getting food. These inherited behaviors provide advantages that promote survival of the species.

Reading Check *Why is social behavior important?*

Figure 9
When several zebras are close together their stripes make it difficult for predators to pick out one individual.

140 ◆ C CHAPTER 5 Animal Behavior

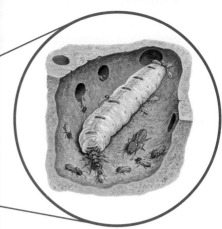

Figure 10 Termites built this large mound in Australia. The mound has a network of tunnels and chambers for the queen to deposit eggs into.

Societies Insects such as ants, bees, and the termites shown in **Figure 10**, live together in societies. A **society** is a group of animals of the same species living and working together in an organized way. Each member has a certain role. Usually a specific female lays eggs, and a male fertilizes them. Workers do all the other jobs in the society.

Some societies are organized by dominance. Wolves usually live together in packs. A wolf pack has a dominant female. The top female controls the mating of the other females. If plenty of food is available, she mates and then allows the others to do so. If food is scarce, she allows less mating. During such times, she is usually the only one to mate.

Territorial Behavior

Many animals set up territories for feeding, mating, and raising young. A territory is an area that an animal defends from other members of the same species. Ownership of a territory occurs in different ways. Songbirds sing, sea lions bellow, and squirrels chatter to claim territories. Other animals leave scent marks. Some animals, like the tiger in **Figure 11,** patrol an area and attack other animals of the same species who enter their territory. Why do animals defend their territories? Territories contain food, shelter, and potential mates. If an animal has a territory, it will be able to mate and produce offspring. Defending territories is an instinctive behavior. It improves the survival rate of an animal's offspring.

Figure 11 A tiger's territory may include several miles. It will confront any other tiger who enters it.

Figure 12
Young wolves roll over and make themselves as small as possible to show their submission to adult wolves.

Figure 13
During the waggle dance, if the source is far from the hive, the dance takes the form of a figure eight. The angle of the waggle is equal to the angle from the hive between the Sun and nectar source.

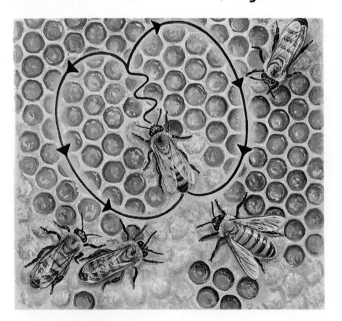

Aggression Have you ever watched as one dog approached another dog that was eating a bone? What happened to the appearance of the dog with the bone? Did its hair on its back stick up? Did it curl its lips and make growling noises? This behavior is aggression. **Aggression** is a forceful behavior used to dominate or control another animal. Fighting and threatening are aggressive behaviors animals use to defend their territories, protect their young, or to get food.

Many animals demonstrate aggression. Some birds let their wings droop below their tail feathers. It may take another bird's perch and thrust its head forward in a pecking motion as a sign of aggression. Cats lay their ears flat, arch their backs, and hiss.

Submission Animals of the same species seldom fight to the death. Teeth, beaks, claws, and horns are used for killing prey or for defending against members of a different species.

To avoid being attacked and injured by an individual of its own species, an animal shows submission. Postures that make an animal appear smaller often are used to communicate surrender. In some animal groups, one individual is usually dominant. Members of the group show submissive behavior toward the dominant individual. This stops further aggressive behavior by the dominant animal. Young animals also display submissive behaviors toward parents or dominant animals, as shown in **Figure 12**.

Communication

In all social behavior, communication is important. Communication is an action by a sender that influences the behavior of a receiver. How do you communicate with the people around you? You may talk, make noises, or gesture like you did in this chapter's Explore Activity. Honeybees perform a dance, as shown in **Figure 13**, to communicate to other bees in the hive where a food source is. Animals in a group communicate with sounds, scents, and actions. Alarm calls, chemicals, speech, courtship behavior, and aggression are forms of communication.

Figure 14
This male Emperor of Germany bird of paradise attracts mates by posturing and fanning its tail.

Courtship Behavior A male bird of paradise, shown in **Figure 14,** spreads its tail feathers and struts. A male sage grouse fans its tail, fluffs its feathers, and blows up its two red air sacs. These are examples of behavior that animals perform before mating. This type of behavior is called **courtship behavior.** Courtship behaviors allow male and female members of a species to recognize each other. These behaviors also stimulate males and females so they are ready to mate at the same time. This helps ensure reproductive success.

In most species the males are more colorful and perform courtship displays to attract a mate. Some courtship behaviors allow males and females to find each other across distances.

Chemical Communication

Ants are sometimes seen moving single file toward a piece of food. Male dogs frequently urinate on objects and plants. Both behaviors are based on chemical communication. The ants have laid down chemical trails that others of their species can follow. The dog is letting other dogs know he has been there. In these behaviors, the animals are using a chemical called a pheromone to communicate. A **pheromone** (FER uh mohn) is a chemical that is produced by one animal to influence the behavior of another animal of the same species. They are powerful chemicals needed only in small amounts. They remain in the environment so that the sender and the receiver can communicate without being in the same place at the same time. They can advertise the presence of an animal to predators, as well as to the intended receiver of the message.

Males and females use pheromones to establish territories, warn of danger, and attract mates. Certain ants, mice, and snails release alarm pheromones when injured or threatened.

Demonstrating Chemical Communication

Procedure
1. Obtain a **sample of perfume or air freshener.**
2. Spray it into the air to leave a scent trail as you move around the house or apartment to a hiding place.
3. Have someone try to discover where you are by following the scent of the substance.

Analysis
1. What was the difference between the first and last room you were in?
2. Would this be an efficient way for humans to communicate? Explain.

SECTION 2 Behavioral Interactions **C ◆ 143**

Figure 15
Many animals use sound to communicate.

A Tree frogs often croak loud enough to be heard far away.

B Pileated woodpecker calls often can be heard above everything else in the forest.

C Howler monkeys got their name because of the sounds they make.

Sound Communication Male crickets rub one forewing against the other forewing. This produces chirping sounds that attract females. Each cricket species produces several calls that are different from other cricket species. These calls are used by researchers to identify different species. Male mosquitoes have hairs on their antennae that sense buzzing sounds produced by females of their same species. The tiny hairs vibrate only to the frequency emitted by a female of the same species.

Vertebrates use a number of different forms of sound communication. Rabbits thump the ground, gorillas pound their chests, beavers slap the water with their flat tails, and frogs, like the one in **Figure 15,** croak. Do you think that sound communication in noisy environments is useful? Seabirds that live where waves pound the shore rather than in some quieter place must rely on visual signals, not sound, for communication.

Chemistry
INTEGRATION

The light produced by fireflies is a particle of visible light that radiates when chemicals produce a high-energy state and then return to their normal state. Hypothesize how this helps fireflies survive. Write your hypothesis in your Science Journal.

Light Communication Certain kinds of flies, marine organisms, and beetles have a special form of communication called bioluminescence. Bioluminescence, shown in **Figure 16,** is the ability of certain living things to give off light. This light is produced through a series of chemical reactions in the organism's body. Probably the most familiar bioluminescent organisms in North America are fireflies. They are not flies, but beetles. The flash of light is produced on the underside of the last abdominal segments and is used to locate a prospective mate. Each species has its own characteristic flashing. Males fly close to the ground and emit flashes of light. Females must flash an answer at exactly the correct time to attract males.

NATIONAL GEOGRAPHIC VISUALIZING BIOLUMINESCENCE

Figure 16

Many marine organisms use bioluminescence as a form of communication. This visible light is produced by a chemical reaction and often confuses predators or attracts mates. Each organism on this page is shown in its normal and bioluminescent state.

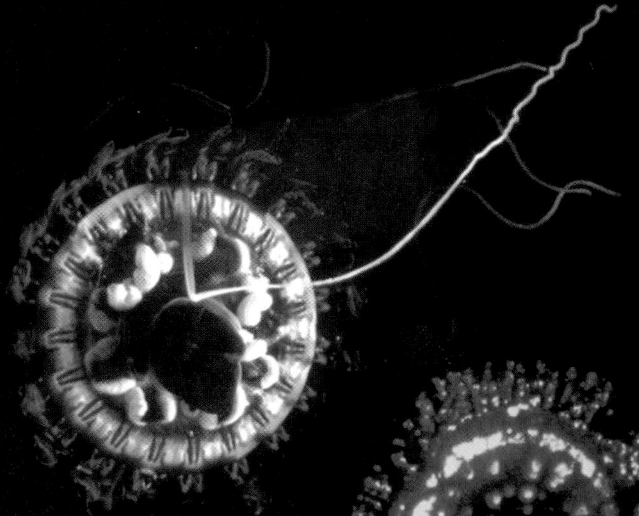

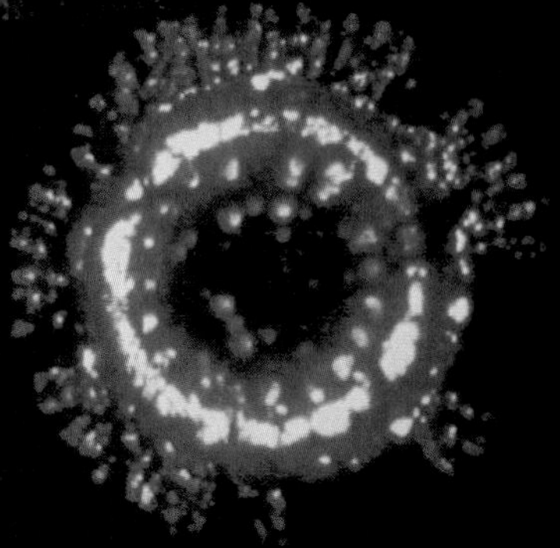

▼ **KRILL** The blue dots shown below this krill are all that are visible when krill bioluminesce. The krill may use bioluminescence to confuse predators.

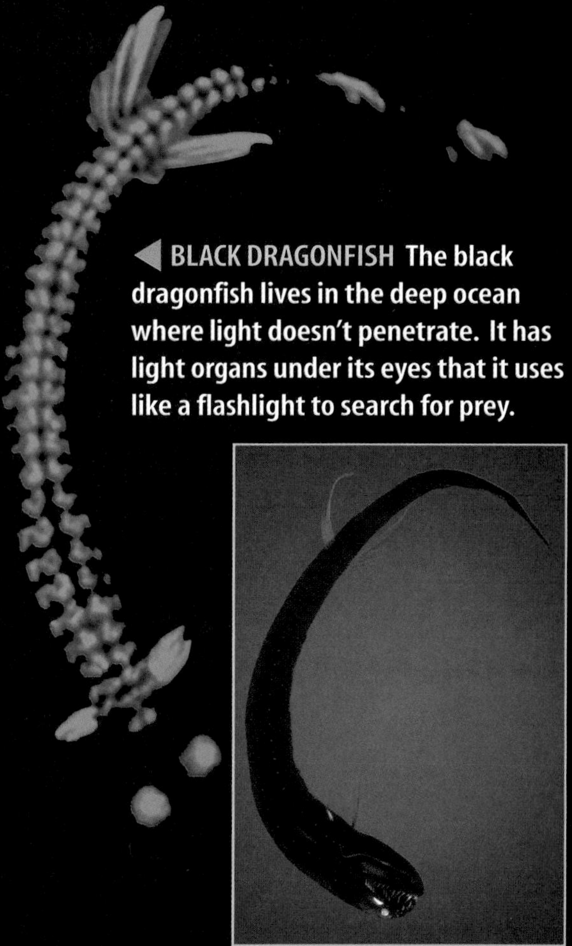

▲ **JELLYFISH** This jellyfish lights up like a neon sign when it is threatened.

◄ **BLACK DRAGONFISH** The black dragonfish lives in the deep ocean where light doesn't penetrate. It has light organs under its eyes that it uses like a flashlight to search for prey.

▲ **DEEP-SEA SEA STAR** The sea star uses light to warn predators of its unpleasant taste.

Uses of Bioluminescence Many bioluminescent animals are found deep in oceans where sunlight does not reach. The ability to produce light may serve several functions. One species of fish dangles a special luminescent organ in front of its mouth. This lures prey close enough to be caught and eaten. Deep-sea shrimp secrete clouds of a luminescent substance when disturbed. This helps them escape their predators. Patterns of luminescence on an animal's body may serve as marks of recognition similar to the color patterns of animals that live in sunlit areas.

Cyclic Behavior

Why do most songbirds rest at night while some species of owls rest during the day? Some animals like the owl in **Figure 17** show regularly repeated behaviors such as sleeping in the day and feeding at night.

A **cyclic behavior** is innate behavior that occurs in a repeating pattern. It often is repeated in response to changes in the environment. Behavior that is based on a 24-hour cycle is called a circadian rhythm. Most animals come close to this 24-hour cycle of sleeping and wakefulness. Experiments have shown that even if animals can't tell whether it was night or day, they continue to behave in a 24-hour cycle.

Animals that are active during the day are diurnal (dy UR nul). Animals that are active at night are nocturnal. Owls are nocturnal. They have round heads, big eyes, and flat faces. Their flat faces reflect light and help them see at night. Owls also have soft feathers that make them almost silent while flying.

Reading Check *What is a diurnal behavior?*

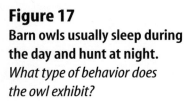

Research Visit the Glencoe Science Web site at **science.glencoe.com** for more information about owl behavior. Communicate to your class what you learn.

Figure 17
Barn owls usually sleep during the day and hunt at night.
What type of behavior does the owl exhibit?

Hibernation Some cyclic behaviors also occur over long periods of time. **Hibernation** is a cyclic response to cold temperatures and limited food supplies. During hibernation, an animal's body temperature drops to near that of its surroundings, and its breathing rate is greatly reduced. Animals in hibernation, such as the bats in **Figure 18,** survive on stored body fat. The animal remains inactive until the weather becomes warm in the spring. Some mammals and many amphibians and reptiles hibernate.

Animals that live in desert like environments also go into a state of reduced activity. This period of inactivity is called estivation. Desert animals sometimes estivate due to extreme heat, lack of food, or periods of drought.

Figure 18
Many bats find a frost-free place like this abandoned coal mine to hibernate for the winter when food supplies are low.

Problem-Solving Activity

How can you determine which animals hibernate?

Many animals hibernate in the winter. During this period of inactivity, they survive on stored body fat. While they are hibernating, they undergo several physical changes. Heart rate slows down and body temperature decreases. The degree to which the body temperature decreases varies among animals. Scientists have disagreed about whether some animals truly hibernate or if they just reduce their activity and go into a light sleep. Usually, a true hibernator's body temperature will decrease significantly while it is hibernating.

Identifying the Problem

The table on the right shows the difference between the normal body temperature and the hibernating body temperature of several animals. What similarities do you notice?

Average Body Temperatures of Hibernating Animals		
Animal	Normal Body Temperature (°C)	Hibernating Body Temperature (°C)
Woodchuck	37	3
Squirrel	32	4
Grizzly Bear	32–37	27–32
Whippoorwill	40	18
Hoary Marmot	37	10

Solving the Problem

1. Which animals would you classify as true hibernators and which would you classify as light sleepers? Explain.
2. Some animals such as snakes and frogs also hibernate. Why would it be difficult to record their normal body temperature on this table?
3. Which animal has the least amount of change in body temperature?

SECTION 2 Behavioral Interactions

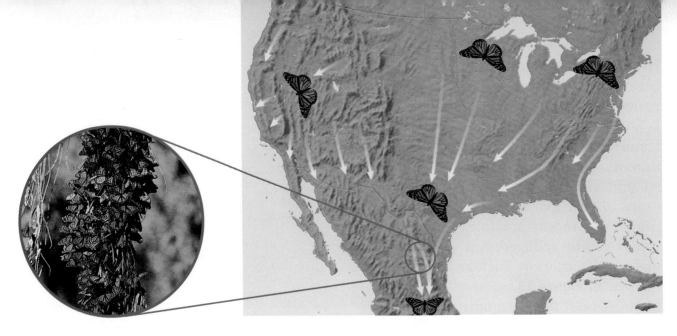

Figure 19
Many monarch butterflies travel from the United States to Mexico for the winter.

Migration Instead of hibernating, many birds and mammals move to new locations when the seasons change. This instinctive seasonal movement of animals is called **migration.** Most animals migrate to find food or reproduce in an environment that is more favorable for the survival of its offspring. Many species of birds fly for hours or days without stopping. The blackpoll warbler flies more than 4,000 km nonstop from North America to its winter home in South America. The trip takes nearly 90 hours. Monarch butterflies, shown in **Figure 19,** can migrate as much as 5,000 km. Gray whales swim from cold arctic waters to the waters off the coast of northern Mexico. After the young are born, they make the return trip.

Section 2 Assessment

1. What are some examples of courtship behavior? How does this behavior help organisms survive?
2. How are cyclic behaviors, such as hibernation, a response to stimuli in the environment?
3. Give two reasons why animals migrate.
4. What is the difference between hibernation and migration?
5. **Think Critically** Suppose a species of frog lives close to a loud waterfall. It often waves a bright blue foot in the air. What might the frog be doing?

Skill Builder Activities

6. **Testing a Hypothesis** Design an experiment that tests the hypothesis that ants leave chemical trails to show other ants where food can be found. **For more help, refer to the Science Skill Handbook.**
7. **Solving One-Step Equations** Some cicadas emerge from the ground every 17 years. The population of one type of caterpillar peaks every five years. If the peak cycle of the caterpillars and the emergence of cicadas coincided in 1990, in what year will they coincide again? **For more help, refer to the Math Skill Handbook.**

Activity

Observing Earthworm Behavior

Earthworms often can be seen wriggling across sidewalks, driveways, and yards on moist nights. Why don't you see many earthworms during the day?

What You'll Investigate
How do earthworms respond to light?

Materials
scissors
shoe box with lid
flashlight
tape
paper
moist paper towels
earthworms
timer

Goals
- **Predict** how earthworms will behave in the presence of light.

Safety Precautions

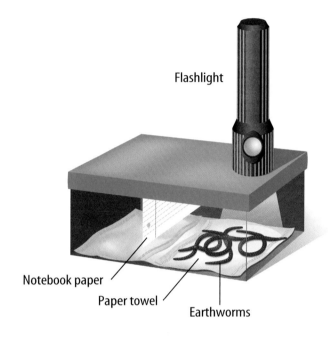

Procedure
1. Cut a round hole, smaller than the end of the flashlight, near one end of the lid.
2. Tape a sheet of paper to the lid so it hangs just above the bottom of the box and about 10 cm away from the end with the hole in it.
3. Place the moist paper towels in the bottom of the box.
4. Place the earthworms in the end of the box that has the hole in it.
5. Hold the flashlight over the hole and turn it on.
6. Leave the box undisturbed for 30 minutes, then open the lid and observe the worms.
7. **Record** the results of your experiment in your Science Journal.

Conclude and Apply
1. Which direction did the earthworms move when the light was turned on?
2. Based on your observations, what can you infer about earthworms?
3. What type of behavior did the earthworms exhibit? Explain.
4. **Predict** where you would need to go to find earthworms during the day.

Communicating Your Data
Write a story that describes a day in the life of an earthworm. List activities, dangers, and problems an earthworm can face. Include a description of its habitat. **For more help, refer to the** Science Skill Handbook.

Activity: Model and Invent

Animal Habitats

Zoos, animal parks, and aquariums are safe places for wild animals. Years ago, captive animals were kept in small cages or behind glass windows. Almost no attempt was made to provide natural habitats for the animals. People who came to see the animals could not observe the animal's normal behavior. Now, most captive animals are kept in exhibit areas that closely resemble their natural habitats. These areas provide suitable environments for the animals so that they can interact with members of their same species and have healthier, longer lives.

Recognize the Problem

What types of environments are best suited for raising animals in captivity?

Thinking Critically

How can the habitats provided at an animal park affect the behavior of animals?

Goals

- **Research** the natural habitat and basic needs of one animal.
- **Design** and model an appropriate zoo, animal park, or aquarium environment for this animal. Working cooperatively with your classmates, design an entire zoo or animal park.

Possible Materials

poster board
markers or colored pencils
materials that can be used to make a scale model

Data Source

SCIENCE *Online* Go to the Glencoe Science Web site at **science.glencoe.com** for more information about existing zoos, animal parks, and aquariums.

Using Scientific Methods

Planning the Model

1. **Choose** an animal to research. Find out where this animal is found in nature. What does it eat? What are its natural predators? Does it exhibit unique territorial, courtship, or other types of behavior? How is this animal adapted to its natural environment?

2. **Design** a model of a proposed habitat in which this animal can live successfully. Don't forget to include all of the things, such as shelter, food, and water, that your animal will need to survive. Will there be any other organisms in the habitat?

Check the Model Plans

1. **Present** your design to your class in the form of a poster, slide show, or video. Compare your proposed habitat with that of the animal's natural environment. Make sure you include a picture of your animal in its natural environment.

2. **Research** how zoos, animal parks, or aquariums provide habitats for animals. Information may be obtained by viewing the Glencoe Science Web site and contacting scientists who work at zoos, animal parks, and aquariums.

Making the Model

1. Using all of the information you have gathered, create a model exhibit area for your animal.

2. Indicate what other plants and animals may be present in the exhibit area.

Analyzing and Applying Results

1. **Decide** whether all of the animals studied in this activity can coexist in the same zoo or wildlife preserve.

2. **Predict** which animals could be grouped together in exhibit areas.

3. **Determine** how large your zoo or wildlife preserve needs to be. Which animals require a large habitat?

4. Using the information provided by the rest of your classmates, design an entire zoo or aquarium that could include the majority of animals studied.

5. **Analyze** problems that might exist in your design. Suggest some ways you might want to improve your design.

Communicating Your Data

Give an oral presentation to another class on the importance of providing natural habitats for captive animals. **For more help, refer to the** Science Skill Handbook.

Oops! Accidents in SCIENCE
SOMETIMES GREAT DISCOVERIES HAPPEN BY ACCIDENT!

Going to the Dogs

A simple and surprising stroll showed that dogs really are humans' best friends

German shepherds make excellent guide dogs.

You've probably seen visually impaired people walking with their trusted and gentle four-legged guides—or "seeing-eye" dogs. The specially-trained dogs serve as eyes for people who can't see, making it possible for them to lead independent lives. But what you probably didn't know is that about 80 years ago, a doctor and his patient discovered this canine ability entirely by accident!

Many people were killed or injured during World War I. Near the end of that war, Dr. Gerhard Stalling and his dog strolled with a patient—a German soldier who had been blinded—around hospital grounds in Germany.

A dog safely guides its owner across a street.

While they were walking, the doctor was briefly called away. The dog and the soldier stayed outside. A few moments later, when the doctor returned, the dog and the soldier were gone! Searching the paths frantically, Dr. Stalling made an astonishing discovery. His pet had led the soldier safely around the hospital grounds. And together the two strolled peacefully back toward the doctor.

School for Dogs

Inspired by what his dog could do, Dr. Stalling set up the first school in the world dedicated to training dogs as guides. Dorothy Eustis, an American woman working as a dog trainer for the International Red Cross in Switzerland, traveled to Stalling's school about ten years later. A report of her visit and study of the way Stalling trained dogs appeared in a New York City newspaper in 1927.

Hearing the story, Morris Frank, a visually impaired American, became determined to get himself a guide dog. He wrote to Dorothy Eustis and asked that she train a dog for him. She accepted his request on one condition.

She wanted Frank to join her in Switzerland for the training process. Frank and his guide dog Buddy returned to New Jersey in 1928. Within a year, Frank set up a training facility in New Jersey, "The Seeing Eye, Inc."

German shepherds, golden retrievers, and Labrador retrievers seem to make the best guide dogs. They learn hand gestures and simple commands to lead visually impaired people across streets and safely around obstacles. This is what scientists call "learned behavior." Animals gain learned behavior through experience. Learning happens gradually and in steps. In fact, scientists say that learning is a somewhat permanent change in behavior due to experience. But, a guide dog not only learns to respond to special commands, it must also know when *not* to obey. If its human owner urges the dog to cross the street and the dog sees that a car is approaching and refuses, the dog has learned to disobey the command. This trait, called "intelligent disobedience," ensures the safety of the owner and the dog—a sure sign that dogs are still humans' best friends.

This girl gets to help train a future guide dog for The Seeing Eye, Inc.

CONNECTIONS Write Lead a blindfolded partner around the classroom. Help your partner avoid obstacles. Then trade places. Write in your Science Journal about your experience leading and being led.

SCIENCE *Online*
For more information, visit
science.glencoe.com

Chapter 5 Study Guide

Reviewing Main Ideas

Section 1 Types of Behavior

1. Behavior that an animal has when it's born is innate behavior. Other animal behaviors are learned through experience. *In the figure below, what type of behavior is the dog exhibiting?*

2. Reflexes are simple innate behaviors. An instinct is a complex pattern of innate behavior.

3. Learned behavior includes imprinting, in which an animal forms a social attachment immediately after birth.

4. Behavior modified by experience is learning by trial and error.

5. Conditioning occurs when the response to one stimulus becomes associated with another. Insight uses past experiences to solve new problems.

Section 2 Behavioral Interactions

1. Behavioral adaptations such as defense of territory, courtship behavior, and social behavior help species of animals survive and reproduce.

2. Courtship behaviors allow males and females to recognize each other and prepare to mate.

3. Interactions among members of the same species are social behaviors. *What type of social behavior is this male peacock displaying?*

4. Communication among organisms occurs in several ways including chemical, sound, and light. *How will other ants, like the one shown, be able to locate food that is far from their nest?*

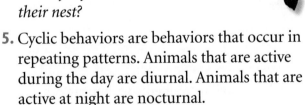

5. Cyclic behaviors are behaviors that occur in repeating patterns. Animals that are active during the day are diurnal. Animals that are active at night are nocturnal.

After You Read

Compare and contrast the behaviors of Animal 1 and Animal 2 listed in your foldable. How many of the behaviors you listed were innate? Learned?

Chapter 5 Study Guide

Visualizing Main Ideas

Complete the following concept map on types of behavior.

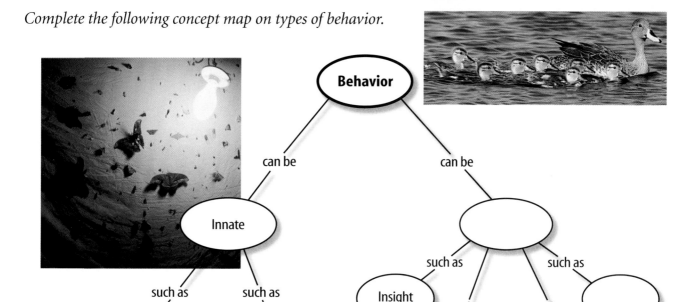

Vocabulary Review

Vocabulary Words

a. aggression
b. behavior
c. conditioning
d. courtship behavior
e. cyclic behavior
f. hibernation
g. imprinting
h. innate behavior
i. insight
j. instinct
k. migration
l. pheromone
m. reflex
n. social behavior
o. society

Using Vocabulary

Explain the differences between the vocabulary words given below. Then explain how the words are related.

1. conditioning, imprinting
2. innate behavior, social behavior
3. insight, instinct
4. social behavior, society
5. instinct, reflex
6. hibernation, migration
7. courtship behavior, pheromone
8. cyclic behavior, migration
9. aggression, social behavior
10. behavior, reflex

Study Tip

Take good notes, even during lab. Lab experiments reinforce key concepts, and looking back on these notes can help you better understand what happened and why.

Chapter 5 Assessment

Checking Concepts

Choose the word or phrase that best answers the question.

1. What is an instinct an example of?
    A) innate behavior   C) imprinting
    B) learned behavior   D) conditioning

2. What is a spider spinning a web an example of?
    A) conditioning   C) learned behavior
    B) imprinting   D) an instinct

3. Which animals depend least on instinct and most on learning?
    A) birds   C) mammals
    B) fish   D) amphibians

4. What is an area that an animal defends from other members of the same species called?
    A) society   C) migration
    B) territory   D) aggression

5. What is a forceful act used to dominate or control?
    A) courtship   C) aggression
    B) reflex   D) hibernation

6. Which of the following is NOT an example of courtship behavior?
    A) fluffing feathers
    B) taking over a perch
    C) singing songs
    D) releasing pheromones

7. What is an organized group of animals doing specific jobs called?
    A) community   C) society
    B) territory   D) circadian rhythm

8. What is the response of inactivity and slowed metabolism that occurs during cold conditions?
    A) hibernation   C) migration
    B) imprinting   D) circadian rhythm

9. Which of the following is a reflex?
    A) writing   C) sneezing
    B) talking   D) riding a bicycle

10. What are behaviors that occur in repeated patterns called?
    A) cyclic   C) reflex
    B) imprinting   D) society

Thinking Critically

11. Explain the type of behavior involved when the bell rings at the end of class.

12. Discuss the advantages and disadvantages of migration as a means of survival.

13. Explain how a habit such as tying your shoes, is different from a reflex.

14. Use one example to explain how behavior increases an animal's chance for survival.

15. Hens lay more eggs in the spring when the number of daylight hours increases. How can farmers use this knowledge of behavior to their advantage?

Developing Skills

16. **Testing a Hypothesis** Design an experiment to test a hypothesis about a specific response to a stimulus from an animal.

17. **Recording Observations** Make observations of a dog, cat, or bird for a week. Record what you see. How did the animal communicate with other animals and with you?

Chapter 5 Assessment

18. Forming a Hypothesis Make a hypothesis about how frogs communicate with each other. How could you test your hypothesis?

19. Classifying Make a list of 25 things that you do regularly. Classify each as an innate or learned behavior. Which behaviors do you have more of?

20. Concept Mapping Complete the following concept map about communication. Use these words: *light, sound, chirping, bioluminescence,* and *buzzing.*

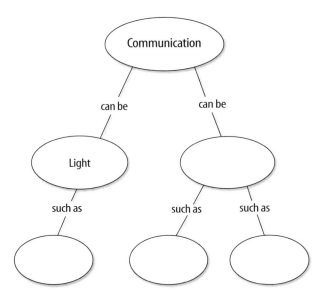

Performance Assessment

21. Poster Draw a map showing the migration route of monarch butterflies, gray whales, or blackpoll warblers.

Technology

Go to the Glencoe Science Web site at **science.glencoe.com** or use the **Glencoe Science CD-ROM** for additional chapter assessment.

 **Test Practice**

A biologist is given illustrations of different behaviors. The different types of behaviors are listed below.

Types of Behavior	
Behavior	Example
1	
2	
3	
4	

Study the table and answer the following questions.

1. A reflex is an automatic response to a stimulus. Which one of the behaviors in the table is an example of a reflex?

   A) one   C) three
   B) two   D) four

2. Trial and error is a type of learned behavior that is modified by experience. Which of the behaviors in the table is an example of a trial-and-error behavior?

   F) one   H) three
   G) two   J) four

Standardized Test Practice

Reading Comprehension

Read the passage. Then read each question that follows the passage. Decide which is the best answer to each question.

Birds and DDT

In the early 1960s, ecologists began to notice a decrease in many bird populations, particularly in the bald eagle population. Ecologists suspected that something must have changed the bald eagles' environments. They worked diligently to discover the cause of the environmental change before the bald eagle was driven to extinction.

Some ecologists thought that the birds were being hurt by something humans were doing. They looked for ways that human activities could be harming the birds. At the time, a pesticide called DDT was being used by farmers to protect their crops from pests. Ecologists realized that after farmers sprayed DDT, it was traveling into lakes and rivers and eventually was entering the groundwater. Once it was in lakes and rivers, aquatic organisms took in the DDT. Fish also became contaminated when they ate DDT-carrying organisms. After eating many DDT-carrying organisms, some fish carried a lot of DDT in their bodies. Then, birds such as the bald eagle ate these fish. Since the birds needed to eat many fish, the birds ended up carrying high levels of DDT in their bodies as well.

While studying the effects of DDT, ecologists discovered that DDT-carrying birds laid eggs with very thin shells. When the birds sat atop their eggs, they inadvertently broke them. As a result, fewer and fewer birds hatched.

Through much hard work and careful research, ecologists were able to determine what was happening to the bird populations.

As a result of their research, DDT use was banned. After several years, the levels of DDT decreased in bird habitats, and many of the bird populations returned to normal levels.

Test-Taking Tip Read the passage slowly and carefully to make sure you understand all the important details.

This drawing shows the bald eagle's food chain.

1. From the story, you can tell that when something is banned, it is _____.
   A) coated with a strip of paint
   B) used to feed baby birds
   C) made illegal
   D) used more often

2. What is the main idea of the second paragraph of this passage?
   F) Once in lakes and rivers, aquatic organisms took in the DDT.
   G) Ecologists discovered that DDT was in the food chain of birds, including the bald eagle.
   H) Fish became contaminated when they ate DDT-carrying organisms.
   J) At the time, a pesticide called DDT was being used by farmers.

Standardized Test Practice

Reasoning and Skills

Read each question and choose the best answer.

Group A

Group B

1. The animals in Group A are different from the animals in Group B because only the animals in Group A _____.
   A) live under water
   B) reproduce asexually
   C) feed by filtering water
   D) reproduce by budding

Test-Taking Tip Think about the different characteristics of sponges and cnidarians.

JAWLESS
CARTILAGE
SCALELESS

2. Which of the following animals have all of the characteristics that are listed above?
   F) shark
   G) tuna
   H) lamprey eel
   J) goldfish

Test-Taking Tip Review the three classes of fish: bony, jawless, and cartilaginous.

Campers dig a hole to bury their food scraps. Before they leave, they refill the hole with soil and place several large rocks on top. Later, raccoons explore the campground area, sniffing the ground. Eventually, the animals dig around and under the rocks to get to the food scraps.

3. If the racoons do not find the food scraps, what eventually will happen to them?
   A) They will remain unchanged.
   B) Earthworms and other soil animals can break them down.
   C) The food scraps will become fungi.
   D) Other campers can reuse them.

4. The above is an example of how racoons _____.
   F) can be affected by pheromones
   G) use insight
   H) show cyclic behavior
   J) prepare for hibernation

Test-Taking Tip Think about the definition of each type of animal behavior listed.

Consider this question carefully before writing your answer on a separate sheet of paper.

5. Recall what you know about animal behavior. Explain why most of the organisms that use bioluminescence for communication live in oceans.

Test-Taking Tip Think about the ocean environment before you answer the question.

Student Resources

Student Resources

CONTENTS

Field Guides — 162

Insects Field Guide 162
Feline Traits Field Guide 166

Skill Handbooks — 170

Science Skill Handbook 170
  Organizing Information 170
    Researching Information 170
    Evaluating Print and
      Nonprint Sources 170
    Interpreting Scientific Illustrations .. 171
    Concept Mapping 171
    Venn Diagram 171
    Writing a Paper 173
  Investigating and Experimenting 174
    Identifying a Question 174
    Forming Hypotheses 174
    Predicting 174
    Testing a Hypothesis 174
    Identifying and Manipulating
      Variables and Controls 175
    Collecting Data 176
    Measuring in SI 177
    Making and Using Tables 178
    Recording Data 179
    Recording Observations 179
    Making Models 179
    Making and Using Graphs 180
  Analyzing and Applying Results 181
    Analyzing Results 181
    Forming Operational Definitions ... 181
    Classifying 181
    Comparing and Contrasting 182
    Recognizing Cause and Effect 182
    Interpreting Data 182
    Drawing Conclusions 182
    Evaluating Other's Data
      and Conclusions 183
    Communicating 183

Technology Skill Handbook 184
  Computer Skills 184
  Using a Word Processor 184
  Using a Database 184
  Using an Electronic Spreadsheet 185
  Using a Computerized Card Catalog ... 186
  Using Graphics Software 186
  Developing Multimedia Presentations . 187

Math Skill Handbook 188
  Converting Units 188
  Using Fractions 189
  Calculating Ratios 190
  Using Decimals 190
  Using Percentages 191
  Using Precision and Significant Digits . 191
  Solving One-Step Equations 192
  Using Proportions 193
  Using Statistics 194

Reference Handbook — 195

A. Safety in the Science Classroom 195
B. SI Metric to English,
   English to Metric Conversions 196
C. Care and Use of a Microscope 197
D. Diversity of Life 198

English Glossary 202
Spanish Glossary 207
Index 212

Field Guide

It's brown and creepy, and it has wings and six legs. If you call it a bug, you might be correct, but if you call it an insect, you are definitely correct. Insects belong to a large group of animals called arthropods. They are related to shrimp, spiders, lobsters, and centipedes. More insect species exist than all other animal species on Earth. Insects are found from the tropics to the tundra. Some live in water all or part of their lives, and some insects even live inside other animals. Insects play important roles in the environment. Many are helpful, but others are destructive.

How Insects Are Classified

An insect's body is divided into three parts—head, thorax, and abdomen. The head has a pair of antennae and eyes and paired mouthparts. Three pairs of jointed legs and, sometimes, wings are attached to the thorax. The abdomen has neither wings nor legs. Insects have a hard covering over their entire body. They shed this covering, then replace it as they grow. Insects are classified into smaller groups called orders. By observing an insect and recognizing certain features, you can identify the order it belongs to. This field guide presents ten insect orders.

Insects

Insect Orders

Convergent ladybug beetle

Coleoptera

Beetles

This is the largest order of insects. Many sizes, shapes, and colors of beetles can be found. All beetles have a pair of thick, leathery wings that meet in a straight line and cover another pair of wings, the thorax, and all or most of the abdomen. Most beetles are considered to be serious pests, but some feed on other insects and others eat dead and decaying organisms. Not all beetles are called beetles. For example, fireflies, June bugs, and weevils are types of beetles.

Male stag beetle

Field Activity

For a week, use this field guide to help identify insect orders. Look for insects in different places and at different times. Visit the Glencoe Science Web site at **science.glencoe.com** to view other insects that might not be found in your city. In your Science Journal, record the order of insect found, along with the date, time, and place.

Dermaptera

Earwigs

The feature that quickly identifies this brown, beetlelike insect is the pair of pincerlike structures that extend from the end of the abdomen. Earwigs usually are active at night and hide under litter or in any dark, protected place during the day. They can damage plants.

Earwig

Common housefly

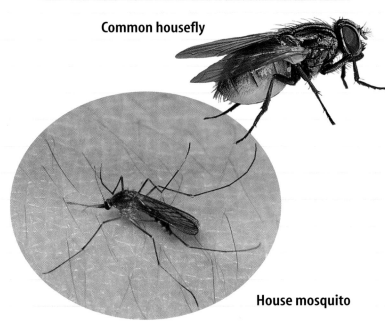

Diptera

Flies and Mosquitoes

These are small insects with large eyes. They have two pair of wings but only one pair can be seen when the insect is at rest and the wings are folded. Their mouths are adapted for piercing and sucking, or scraping and lapping. Many of these insects are food for larger animals. Some spread diseases, others are pests, and some eat dead and decaying organisms. They are found in many different environments.

House mosquito

Odonata

Dragonflies and Damselflies

These insects have two pairs of transparent, multi-veined wings that are nearly equal in size and are never folded against the insect's body. The head has a pair of large eyes on it, and the abdomen is long and thin. These insects are usually seen near bodies of water. All members of this group hunt during flight and catch small insects, such as mosquitoes.

Dragonfly

FIELD GUIDE   C ◆ 163

Isoptera

Termites

Adult termites are small, dark brown or black, and can have wings. Immature forms of this insect are small, soft bodied, pale yellow or white, and wingless. The adults are sometimes confused with ants. The thorax and abdomen of a termite look like one body part, but a thin waist separates the thorax and abdomen of an ant. Termites live in colonies in the ground or in wood.

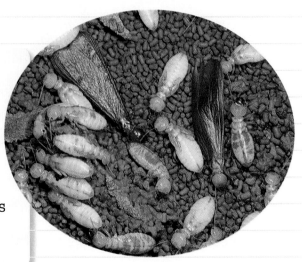

Pacific coast termites

Dictyoptera

Cockroaches and Mantises

These insects have long, thin antennae on the head. The front wings are smaller than the back wings. The back wings are thin and fanlike when they are opened. In the mantis, the front legs are adapted for grasping. The other two pairs of legs are similar to those of a cockroach. Praying mantises are beneficial because they eat other, often harmful, insects. Cockroaches are pests.

American cockroach

Carolina praying mantis

Hymenoptera

Ants, Bees, and Wasps

Members of this order can be so small that they're visible only with a magnifier. Others may be nearly 35 mm long. These insects have two pairs of transparent wings, if present. They are found in many different environments, either in colonies or alone. They are important because they pollinate flowers, and some prey on harmful insects. Honeybees make honey and wax.

Paper wasp

American bumblebee

Black carpenter ant

Lepidoptera

Butterflies and Moths

Butterflies and moths have two pairs of wings with colorful patterns created by thousands of tiny scales. A moth's antennae are feathery. A butterfly's antennae are thin, and each has a small knob on the tip. Adult's mouthparts are adapted as a long, coiled tube for drinking nectar. Moths are active at night, and butterflies are active on warm, sunny days.

Yellow woolly bear moth

Buckeye butterfly

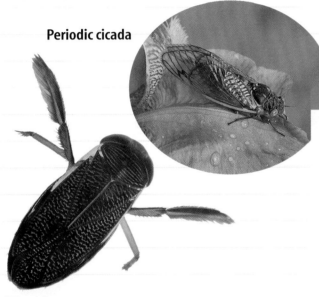

Periodic cicada

Water boatman

Hemiptera

Bugs

The prefix of this order, *"Hemi-"*, means "half" and describes the front pair of wings. Near the insect's head, the front wings are thick and leathery, and are thin at the tip. Wing tips usually overlap when they are folded over the insect's back and cover a smaller pair of thin wings. Some bugs live on land and others are aquatic.

Orthoptera

Grasshoppers, Crickets, and Katydids

These insects have large hind legs adapted for leaping. They usually have two pairs of wings. The outer pair is hard and covers a transparent pair. Many of these insects make singing noises by rubbing one body part against another. Males generally make these sounds. These insects are considered pests because swarms of them can destroy a farmer's crops in a few days.

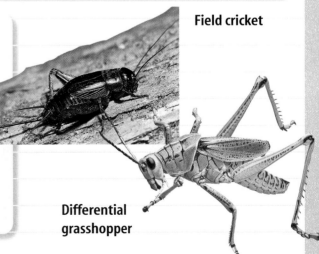

Field cricket

Differential grasshopper

FIELD GUIDE C ◆ 165

Field Guide

Feline Traits

For centuries, humans have lived with cats. They have kept cats in their homes and on their properties. Cats hunt mice and other rodents that eat stored grains and other human foods. Cats also are companions and family pets. Today, the cat is the most popular pet in the United States.

When animals mate, they pass their genetic traits to their offspring. Occasionally, a natural mutation results in a new breed. Sometimes animal scientists try to create new breeds through selective breeding. They study the pedigree, or family history, of several animals of the same species to see what genetic traits they carry. Then they mate the animals that are most likely to pass the desired traits to their offspring. Over time, a new breed can be developed.

The Cat Fancier's Association recognizes 37 breeds of cats. This field guide describes the traits of 14 of these recognized breeds. None of these breeds has come about by recent selective breeding. Some of these breeds have ancient histories, and others have resulted from natural mutations.

Most cats that people adopt today have a mixed ancestry of several breeds. Read about different breeds on the next few pages. Do you know a cat that has one or more of the described traits?

Feline Breeds

Siamese

Siamese

This is an ancient breed from Siam, which is now called Thailand. Siamese have long bodies and tails, and their fur is short. They are easy to recognize because they have light-colored bodies and dark ears, masks, tails, and legs. Their eyes are blue.

Devon Rex

Devon Rex

This breed is a natural mutation that first appeared in Devonshire, England, in 1960. Their eyes and batlike ears look huge against their tiny faces. When you stroke a Devon Rex's coat, its fur ripples.

Field Activity

For a week, use this field guide to observe the cats in your neighborhood. What traits do they have? What breeds might be part of their pedigree? Go to the Glencoe Science Web site at **science.glencoe.com** to find more photographs of felines. In your Science Journal, record each cat's name, the traits you noticed, and the breeds that have those traits.

Cornish Rex

Cornish Rex

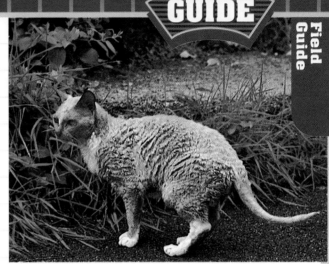

This natural mutation first appeared in Cornwall, England, in about 1950. Cornish Rex cats remind some people of a skinny breed of dog called a whippet. They have arched backs, small waists, and long legs. Their ears are large, and they have small, egg-shaped heads. Their short, curly fur is soft.

Chartreux [shahr-TROOZ]

Chartreux

This breed dates back at least to the sixteenth century. Their name comes from a type of Spanish wool, but they are considered French cats. Chartreux cats are large and powerful, but they tend to be gentle. They are known for their woolly, bluish coats.

British Shorthair

This breed descends from ancient Roman house cats. They are large, powerful animals with broad chests and round faces. They have short legs and short, thick fur. You might have seen these calm, intelligent cats in TV commercials.

British Shorthair

American Wirehair

This breed is a natural mutation that first appeared in New York in 1966. The feature that makes them special is their unusual coat. Each hair is stiff and crimped, which makes the coat hard. Their whiskers also are wiry.

American Wirehair

Field Guide

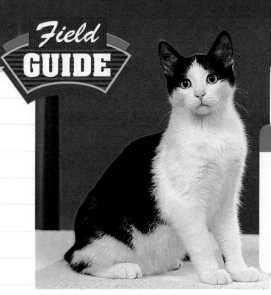

American Shorthair

American Shorthair

This breed came to America with the English Puritans in the 1600s. As the name suggests, they have short fur. They come in many colors, but most are silver with black bands.

Abyssinian (a buh SIH nee un)

In the mid-1800s, this breed was brought to England from Ethiopia, which was then called Abyssinia. However, some scientists believe these cats might have originated in Asia. Abyssinians have arched necks, muscular bodies, large ears, and almond-shaped eyes. Their coats can be ruddy, fawn, auburn, or bluish, and they are marked with several dark bands.

Abyssinian

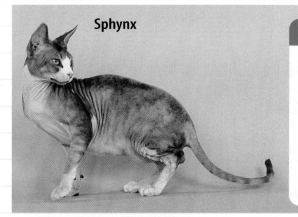

Sphynx

Sphynx

The Sphynx is a natural mutation. It first appeared in Canada in 1966. At first glance, these cats appear to be totally bald. In fact, their bodies are covered with a soft, fuzzy, downlike hair. They have short whiskers or none at all, and their skin is often wrinkled.

Selkirk Rex

This breed first appeared as a natural mutation in Wyoming. Selkirk Rexs are large and powerful like the British Shorthair, and they have curly hair and whiskers. Unlike Devon Rex and Cornish Rex, their hair can be long or short. They come in many colors.

Selkirk Rex

Norwegian Forest Cat

These green-eyed cats came to North America with the Viking explorers. In the winter, their hair is thick and a plush mane grows around their necks and chests. The long hair of their inner ears stays with them all year. They come in many colors.

Norwegian Forest Cat

Manx

This breed first appeared on the Isle of Man. Manx cats are best known as the cats without tails, but some have stubby tails called rises. They have arched backs, round heads, and round bodies. Their hair can be long or short.

Maine Coon

This breed developed in North America and was first recognized in Maine. Maine Coons are large, sturdy cats. Long hair and the tufts of hair in their ears help them tolerate extreme cold. Their coats are shaggy, but they feel silky.

Maine Coon cats

Egyptian Mau

The ancient Egyptians loved to draw this breed, which descends from the African wildcat. Egyptian Maus have green eyes, and their fur can be silver, bronze, black, bluish, or smoke colored. They differ from all other domestic cats because they are the only natural breed with a spotted coat.

Science Skill Handbook

Organizing Information

As you study science, you will make many observations and conduct investigations and experiments. You will also research information that is available from many sources. These activities will involve organizing and recording data. The quality of the data you collect and the way you organize it will determine how well others can understand and use it. In **Figure 1,** the student is obtaining and recording information using a microscope.

Putting your observations in writing is an important way of communicating to others the information you have found and the results of your investigations and experiments.

Researching Information

Scientists work to build on and add to human knowledge of the world. Before moving in a new direction, it is important to gather the information that already is known about a subject. You will look for such information in various reference sources. Follow these steps to research information on a scientific subject:

Step 1 Determine exactly what you need to know about the subject. For instance, you might want to find out what happened to local plant life when Mount St. Helens erupted in 1980.

Step 2 Make a list of questions, such as: When did the eruption begin? How long did it last? How large was the area in which plant life was affected?

Step 3 Use multiple sources such as textbooks, encyclopedias, government documents, professional journals, science magazines, and the Internet.

Step 4 List where you found the sources. Make sure the sources you use are reliable and the most current available.

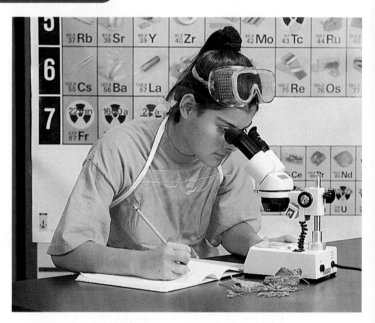

Figure 1
Making an observation is one way to gather information directly.

Evaluating Print and Nonprint Sources

Not all sources of information are reliable. Evaluate the sources you use for information, and use only those you know to be dependable. For example, suppose you want information about the digestion of fats and proteins. You might find two Websites on digestion. One Web site contains "Fat Zapping Tips" written by a company that sells expensive, high-protein supplements to help your body eliminate excess fat. The other is a Web page on "Digestion and Metabolism" written by a well-respected medical school. You would choose the second Web site as the more reliable source of information.

In science, information can change rapidly. Always consult the most current sources. A 1985 source about the human genome would not reflect the most recent research and findings.

Science Skill Handbook

Interpreting Scientific Illustrations

As you research a science topic, you will see drawings, diagrams, and photographs. Illustrations help you understand what you read. Some illustrations are included to help you understand an idea that you can't see easily by yourself. For instance, you can't see the bones of a blue whale, but you can look at a diagram of a whale skeleton as labeled in **Figure 2** that helps you understand them. Visualizing a drawing helps many people remember details more easily. Illustrations also provide examples that clarify difficult concepts or give additional information about the topic you are studying.

Most illustrations have a label or a caption. A label or caption identifies the illustration or provides additional information to better explain it. Can you find the caption or labels in **Figure 2?**

Figure 2
A labeled diagram of the skeletal structure of a blue whale.

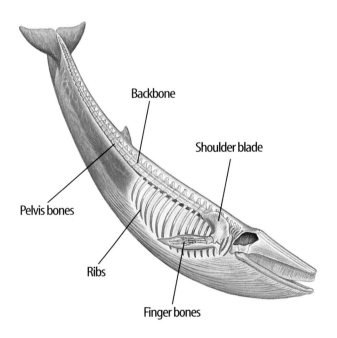

Concept Mapping

If you were taking a car trip, you might take some sort of road map. By using a map, you begin to learn where you are in relation to other places on the map.

A concept map is similar to a road map, but a concept map shows relationships among ideas (or concepts) rather than places. It is a diagram that visually shows how concepts are related. Because a concept map shows relationships among ideas, it can make the meanings of ideas and terms clear and help you understand what you are studying.

Overall, concept maps are useful for breaking large concepts down into smaller parts, making learning easier.

Venn Diagram

Although it is not a concept map, a Venn diagram illustrates how two subjects compare and contrast. In other words, you can see the characteristics that the subjects have in common and those that they do not.

The Venn diagram in **Figure 3** shows the relationship between two categories of organisms, plants and animals. Both share some basic characteristics as living organisms. However, there are differences in the ways they carry out various life processes, such as obtaining nourishment, that distinguish one from the other.

Figure 3
A Venn diagram shows how objects or concepts are alike and how they are different.

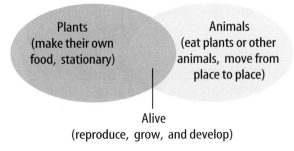

SCIENCE SKILL HANDBOOK C ◆ 171

Science Skill Handbook

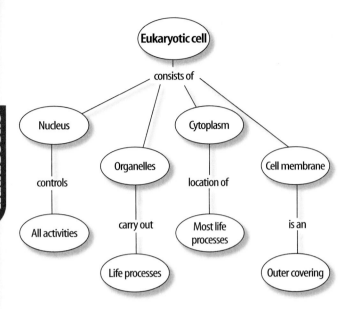

Figure 4
A network tree shows how concepts or objects are related.

Network Tree Look at the concept map in **Figure 4,** that shows details about a eukaryotic cell. This is called a network tree concept map. Notice how some words are in ovals while others are written across connecting lines. The words inside the ovals are science terms or concepts. The words written on the connecting lines describe the relationships between the concepts.

When constructing a network tree, write the topic on a note card or piece of paper. Write the major concepts related to that topic on separate note cards or pieces of paper. Then arrange them in order from general to specific. Branch the related concepts from the major concept and describe the relationships on the connecting lines. Continue branching to more specific concepts. Write the relationships between the concepts on the connecting lines until all concepts are mapped. Then examine the concept map for relationships that cross branches, and add them to the concept map.

Events Chain An events chain is another type of concept map. It models the order of items or their sequence. In science, an events chain can be used to describe a sequence of events, the steps in a procedure, or the stages of a process.

When making an events chain, first find the one event that starts the chain. This event is called the *initiating event*. Then, find the next event in the chain and continue until you reach an outcome. Suppose you are asked to describe the main stages in the growth of a plant from a seed. You might draw an events chain such as the one in **Figure 5.** Notice that connecting words are not necessary in an events chain.

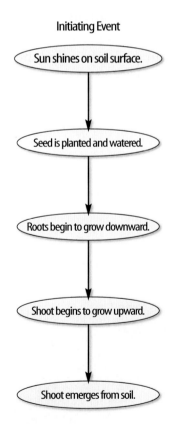

Figure 5
Events chains show the order of steps in a process or event.

172 ◆ C STUDENT RESOURCES

Science Skill Handbook

Cycle Map A cycle concept map is a specific type of events chain map. In a cycle concept map, the series of events does not produce a final outcome. Instead, the last event in the chain relates back to the beginning event.

You first decide what event will be used as the beginning event. Once that is decided, you list events in order that occur after it. Words are written between events that describe what happens from one event to the next. The last event in a cycle concept map relates back to the beginning event. The number of events in a cycle concept varies, but is usually three or more. Look at the cycle map, as shown in **Figure 6**.

Figure 7
A spider map allows you to list ideas that relate to a central topic but not necessarily to one another.

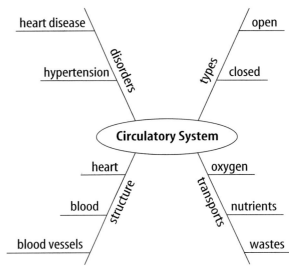

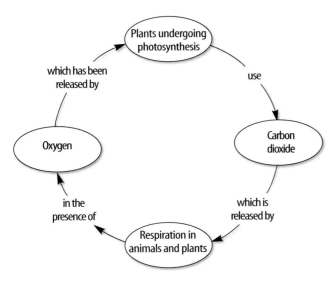

Figure 6
A cycle map shows events that occur in a cycle.

Spider Map A type of concept map that you can use for brainstorming is the spider map. When you have a central idea, you might find you have a jumble of ideas that relate to it but may not clearly relate to each other. The circulatory system spider map in **Figure 7** shows that if you write these ideas outside the main concept, then you can begin to separate and group unrelated terms so they become more useful.

Writing a Paper

You will write papers often when researching science topics or reporting the results of investigations or experiments. Scientists frequently write papers to share their data and conclusions with other scientists and the public. When writing a paper, use these steps.

Step 1 Assemble your data by using graphs, tables, or a concept map. Create an outline.

Step 2 Start with an introduction that contains a clear statement of purpose and what you intend to discuss or prove.

Step 3 Organize the body into paragraphs. Each paragraph should start with a topic sentence, and the remaining sentences in that paragraph should support your point.

Step 4 Position data to help support your points.

Step 5 Summarize the main points and finish with a conclusion statement.

Step 6 Use tables, graphs, charts, and illustrations whenever possible.

SCIENCE SKILL HANDBOOK   C ◆ 173

Science Skill Handbook

Investigating and Experimenting

You might say the work of a scientist is to solve problems. When you decide to find out why one corner of your yard is always soggy, you are problem solving, too. You might observe the corner is lower than the surrounding area and has less vegetation growing in it. You might decide to see if planting some grass will keep the corner drier.

Scientists use orderly approaches to solve problems. The methods scientists use include identifying a question, making observations, forming a hypothesis, testing a hypothesis, analyzing results, and drawing conclusions.

Scientific investigations involve careful observation under controlled conditions. Such observation of an object or a process can suggest new and interesting questions about it. These questions sometimes lead to the formation of a hypothesis. Scientific investigations are designed to test a hypothesis.

Identifying a Question

The first step in a scientific investigation or experiment is to identify a question to be answered or a problem to be solved. You might be interested in knowing why an animal like the one in **Figure 8** look the way they do.

Figure 8
When you see a bird, you might ask yourself, "How does the shape of this bird's beak help it feed?"

Forming Hypotheses

Hypotheses are based on observations that have been made. A hypothesis is a possible explanation based on previous knowledge and observations.

Perhaps a scientist has observed that bean plants grow larger if they are fertilized than if not. Based on these observations, the scientist can make a statement that he or she can test. The statement is a hypothesis. The hypothesis could be: *Fertilizer makes bean plants grow larger.* A hypothesis has to be something you can test by using an investigation. A testable hypothesis is a valid hypothesis.

Predicting

When you apply a hypothesis, or general explanation, to a specific situation, you predict something about that situation. First, you must identify which hypothesis fits the situation you are considering. People use predictions to make everyday decisions. Based on previous observations and experiences, you might form a prediction that if fertilizer makes bean plants grow larger, then fertilized plants will yield more beans than plants not fertilized. Someone could use this prediction to plan to grow fewer plants.

Testing a Hypothesis

To test a hypothesis, you need a procedure. A procedure is the plan you follow in your experiment. A procedure tells you what materials to use, as well as how and in what order to use them. When you follow a procedure, data are generated that support or do not support the original hypothesis statement.

Science Skill Handbook

For example, suppose you notice that your guppies don't seem as active as usual when your aquarium heater is not working. You wonder how water temperature affects guppy activity level. You decide to test the hypothesis, "If water temperature increases, then guppy activity should increase." Then you write the procedure shown in **Figure 9** for your experiment and generate the data presented in the table below.

Procedure
1. Fill five identical glass containers with equal amounts of aquarium water.
2. Measure and record the temperature of the water in the first container.
3. Heat and cool the other containers so that two have higher and two have lower water temperatures.
4. Place a guppy in each container; count and record the number of movements each guppy makes in 5 minutes.

Figure 9
A procedure tells you what to do step by step.

Number of Guppy Movements		
Container	Temperature (°C)	Movements
1	38	56
2	40	61
3	42	70
4	36	46
5	34	42

Are all investigations alike? Keep in mind as you perform investigations in science that a hypothesis can be tested in many ways. Not every investigation makes use of all the ways that are described on these pages, and not all hypotheses are tested by investigations. Scientists encounter many variations in the methods that are used when they perform experiments. The skills in this handbook are here for you to use and practice.

Identifying and Manipulating Variables and Controls

In any experiment, it is important to keep everything the same except for the item you are testing. The one factor you change is called the independent variable. The factor that changes as a result of the independent variable is called the dependent variable. Always make sure you have only one independent variable. If you allow more than one, you will not know what causes the changes you observe in the dependent variable. Many experiments also have controls—individual instances or experimental subjects for which the independent variable is not changed. You can then compare the test results to the control results.

For example, in the guppy experiment, you made everything the same except the temperature of the water. The glass containers were identical. The volume of aquarium water in each container and beginning water temperature were the same. Each guppy was like the others, as much as possible. In this way, you could be sure that any difference in the number of guppy movements was caused by the temperature change—the independent variable. The activity level of the guppy was measured as the number of guppy movements—the dependent variable. The guppy in the container in which the water temperature was not changed was the control.

SCIENCE SKILL HANDBOOK C ◆ 175

Science Skill Handbook

Collecting Data

Whether you are carrying out an investigation or a short observational experiment, you will collect data, or information. Scientists collect data accurately as numbers and descriptions and organize it in specific ways.

Observing Scientists observe items and events, then record what they see. When they use only words to describe an observation, it is called qualitative data. For example, a scientist might describe the color of a bird or the shape of a bird's beak as seen through binoculars. Scientists' observations also can describe how much there is of something. These observations use numbers, as well as words, in the description and are called quantitative data. For example, if a particular dog is described as being "furry, yellow, and short-haired," the data are clearly qualitative. Quantitative data for this dog might include "a mass of 14 kg, a height of 46 cm, and an age of 150 days." Quantitative data often are organized into tables. Then, from information in the table, a graph can be drawn. Graphs can reveal relationships that exist in experimental data.

When you make observations in science, you should examine the entire object or situation first, then look carefully for details. If you're looking at a plant, for instance, check general characteristics such as size and overall structure before using a hand lens to examine the leaves and other smaller structures such as flowers or fruits. Remember to record accurately everything you see.

Scientists try to make careful and accurate observations. When possible, they use instruments such as microscopes, metric rulers, graduated cylinders, thermometers, and balances. Measurements provide numerical data that can be repeated and checked.

Sampling When working with large numbers of objects or a large population, scientists usually cannot observe or study every one of them. Instead, they use a sample or a portion of the total number. To *sample* is to take a small, representative portion of the objects or organisms of a population for research. By making careful observations or manipulating variables within a portion of a group, information is discovered and conclusions are drawn that might apply to the whole population.

Estimating Scientific work also involves estimating. To *estimate* is to make a judgment about the size or the number of something without measuring or counting every object or member of a population. Scientists first count the number of objects in a small sample. Looking through a microscope lens, for example, a scientist can count the number of bacterial colonies in the 1-cm^2 frame shown in **Figure 10.** Then the scientist can multiply that number by the number of cm^2 in the petri dish to get an estimate of the total number of bacterial colonies present.

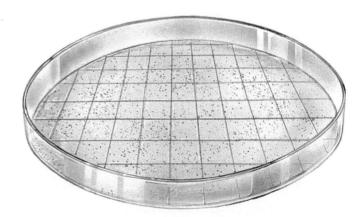

Figure 10
To estimate the total number of bacterial colonies that are present on a petri dish, count the number of bacterial colonies within a 1-cm^2 frame and multiply that number by the number of frames on the dish.

Science Skill Handbook

Measuring in SI

The metric system of measurement was developed in 1795. A modern form of the metric system, called the International System, or SI, was adopted in 1960. SI provides standard measurements that all scientists around the world can understand.

The metric system is convenient because unit sizes vary by multiples of 10. When changing from smaller units to larger units, divide by a multiple of 10. When changing from larger units to smaller, multiply by a multiple of 10. To convert millimeters to centimeters, divide the millimeters by 10. To convert 30 mm to centimeters, divide 30 by 10 (30 mm equal 3 cm).

Prefixes are used to name units. Look at the table below for some common metric prefixes and their meanings. Do you see how the prefix *kilo-* attached to the unit *gram* is *kilogram*, or 1,000 g?

Metric Prefixes			
Prefix	Symbol	Meaning	
kilo-	k	1,000	thousand
hecto-	h	100	hundred
deka-	da	10	ten
deci-	d	0.1	tenth
centi-	c	0.01	hundredth
milli-	m	0.001	thousandth

Now look at the metric ruler shown in **Figure 11.** The centimeter lines are the long, numbered lines, and the shorter lines are millimeter lines.

When using a metric ruler, line up the 0-cm mark with the end of the object being measured, and read the number of the unit where the object ends, in this instance it would be 4.5 cm.

Figure 11
This metric ruler shows centimeters and millimeter divisions.

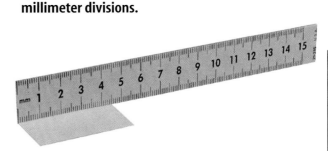

Liquid Volume In some science activities, you will measure liquids. The unit that is used to measure liquids is the liter. A liter has the volume of 1,000 cm^3. The prefix *milli-* means "thousandth (0.001)." A milliliter is one thousandth of 1 L and 1 L has the volume of 1,000 mL. One milliliter of liquid completely fills a cube measuring 1 cm on each side. Therefore, 1 mL equals 1 cm^3.

You will use beakers and graduated cylinders to measure liquid volume. A graduated cylinder, as illustrated in **Figure 12,** is marked from bottom to top in milliliters. This graduated cylinder contains 79 mL of a liquid.

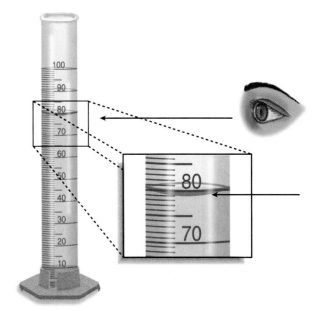

Figure 12
Graduated cylinders measure liquid volume.

Science Skill Handbook

Mass Scientists measure mass in grams. You might use a beam balance similar to the one shown in **Figure 13**. The balance has a pan on one side and a set of beams on the other side. Each beam has a rider that slides on the beam.

Before you find the mass of an object, slide all the riders back to the zero point. Check the pointer on the right to make sure it swings an equal distance above and below the zero point. If the swing is unequal, find and turn the adjusting screw until you have an equal swing.

Place an object on the pan. Slide the largest rider along its beam until the pointer drops below zero. Then move it back one notch. Repeat the process on each beam until the pointer swings an equal distance above and below the zero point. Sum the masses on each beam to find the mass of the object. Move all riders back to zero when finished.

Figure 13
A triple beam balance is used to determine the mass of an object.

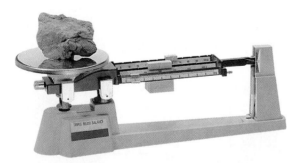

You should never place a hot object on the pan or pour chemicals directly onto the pan. Instead, find the mass of a clean container. Remove the container from the pan, then place the chemicals in the container. Find the mass of the container with the chemicals in it. To find the mass of the chemicals, subtract the mass of the empty container from the mass of the filled container.

Making and Using Tables

Browse through your textbook and you will see tables in the text and in the activities. In a table, data, or information, are arranged so that they are easier to understand. Activity tables help organize the data you collect during an activity so results can be interpreted.

Making Tables To make a table, list the items to be compared in the first column and the characteristics to be compared in the first row. The title should clearly indicate the content of the table, and the column or row heads should tell the reader what information is found in there. The table below lists materials collected for recycling on three weekly pick-up days. The inclusion of kilograms in parentheses also identifies for the reader that the figures are mass units.

Recyclable Materials Collected During Week			
Day of Week	Paper (kg)	Aluminum (kg)	Glass (kg)
Monday	5.0	4.0	12.0
Wednesday	4.0	1.0	10.0
Friday	2.5	2.0	10.0

Using Tables How much paper, in kilograms, is being recycled on Wednesday? Locate the column labeled "Paper (kg)" and the row "Wednesday." The information in the box where the column and row intersect is the answer. Did you answer "4.0"? How much aluminum, in kilograms, is being recycled on Friday? If you answered "2.0," you understand how to read the table. How much glass is collected for recycling each week? Locate the column labeled "Glass (kg)" and add the figures for all three rows. If you answered "32.0," then you know how to locate and use the data provided in the table.

Science Skill Handbook

Recording Data

To be useful, the data you collect must be recorded carefully. Accuracy is key. A well-thought-out experiment includes a way to record procedures, observations, and results accurately. Data tables are one way to organize and record results. Set up the tables you will need ahead of time so you can record the data right away.

Record information properly and neatly. Never put unidentified data on scraps of paper. Instead, data should be written in a notebook like the one in **Figure 14.** Write in pencil so information isn't lost if your data gets wet. At each point in the experiment, record your data and label it. That way, your information will be accurate and you will not have to determine what the figures mean when you look at your notes later.

Figure 14
Record data neatly and clearly so it is easy to understand.

Recording Observations

It is important to record observations accurately and completely. That is why you always should record observations in your notes immediately as you make them. It is easy to miss details or make mistakes when recording results from memory. Do not include your personal thoughts when you record your data. Record only what you observe to eliminate bias. For example, when you record that a plant grew 12 cm in one day, you would note that this was the largest daily growth for the week. However, you would not refer to the data as "the best growth spurt of the week."

Making Models

You can organize the observations and other data you collect and record in many ways. Making models is one way to help you better understand the parts of a structure you have been observing or the way a process for which you have been taking various measurements works.

Models often show things that are very large or small or otherwise would be difficult to see and understand. You can study blood vessels and know that they are hollow tubes. The size and proportional differences among arteries, veins, and capillaries can be explained in words. However, you can better visualize the relative sizes and proportions of blood vessels by making models of them. Gluing different kinds of pasta to thick paper so the openings can be seen can help you see how the differences in size, wall thickness, and shape among types of blood vessels affect their functions.

Other models can be devised on a computer. Some models, such as disease control models used by doctors to predict the spread of the flu, are mathematical and are represented by equations.

Science Skill Handbook

Making and Using Graphs

After scientists organize data in tables, they might display the data in a graph that shows the relationship of one variable to another. A graph makes interpretation and analysis of data easier. Three types of graphs are the line graph, the bar graph, and the circle graph.

Line Graphs A line graph like in **Figure 15** is used to show the relationship between two variables. The variables being compared go on two axes of the graph. For data from an experiment, the independent variable always goes on the horizontal axis, called the *x*-axis. The dependent variable always goes on the vertical axis, called the *y*-axis. After drawing your axes, label each with a scale. Next, plot the data points.

A data point is the intersection of the recorded value of the dependent variable for each tested value of the independent variable. After all the points are plotted, connect them.

Bar Graphs Bar graphs compare data that do not change continuously. Vertical bars show the relationships among data.

To make a bar graph, set up the *y*-axis as you did for the line graph. Draw vertical bars of equal size from the *x*-axis up to the point on the *y*-axis that represents value of *x*.

Figure 16
The number of wing vibrations per second for different insects can be shown as a bar graph or circle graph.

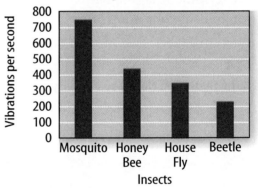

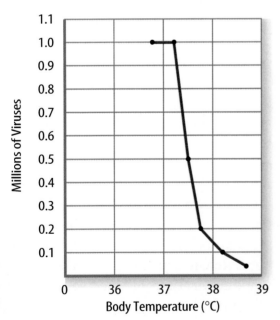

Figure 15
This line graph shows the relationship between body temperature and the millions of infecting viruses present in a human body.

Circle Graphs A circle graph uses a circle divided into sections to display data as parts (fractions or percentages) of a whole. The size of each section corresponds to the fraction or percentage of the data that the section represents. So, the entire circle represents 100 percent, one-half represents 50 percent, one-fifth represents 20 percent, and so on.

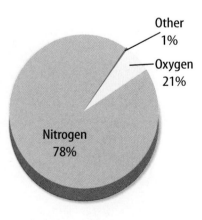

Science Skill Handbook

Analyzing and Applying Results

Analyzing Results

To determine the meaning of your observations and investigation results, you will need to look for patterns in the data. You can organize your information in several of the ways that are discussed in this handbook. Then you must think critically to determine what the data mean. Scientists use several approaches when they analyze the data they have collected and recorded. Each approach is useful for identifying specific patterns in the data.

Forming Operational Definitions

An operational definition defines an object by showing how it functions, works, or behaves. Such definitions are written in terms of how an object works or how it can be used; that is, they describe its job or purpose.

For example, a ruler can be defined as a tool that measures the length of an object (how it can be used). A ruler also can be defined as something that contains a series of marks that can be used as a standard when measuring (how it works).

Classifying

Classifying is the process of sorting objects or events into groups based on common features. When classifying, first observe the objects or events to be classified. Then select one feature that is shared by some members in the group but not by all. Place those members that share that feature into a subgroup. You can classify members into smaller and smaller subgroups based on characteristics.

How might you classify a group of animals? You might first classify them by putting all of the dogs, cats, lizards, snakes, and birds into separate groups. Within each group, you could then look for another common feature by which to further classify members of the group, such as size or color.

Remember that when you classify, you are grouping objects or events for a purpose. For example, classifying animals can be the first step in identifying them. You might know that a cardinal is a red bird. To find it in a large group of animals, you might start with the classification scheme mentioned above. You'll locate a cardinal within the red grouping of the birds that you separate from the rest of the animals. A male ruby-throated hummingbird could be located within the birds by its tiny size and the bright red color of its throat. Keep your purpose in mind as you select the features to form groups and subgroups.

Figure 17
Color is one of many characteristics that are used to classify animals.

SCIENCE SKILL HANDBOOK   C ◆ 181

Science Skill Handbook

Comparing and Contrasting

Observations can be analyzed by noting the similarities and differences between two or more objects or events that you observe. When you look at objects or events to see how they are similar, you are comparing them. Contrasting is looking for differences in objects or events. The table below compares and contrasts the nutritional value of two cereals.

Nutritional Values		
	Cereal A	Cereal B
Calories	220	160
Fat	10 g	10 g
Protein	2.5 g	2.6 g
Carbohydrate	30 g	15 g

Recognizing Cause and Effect

Have you ever gotten a cold and then suggested that you probably caught it from a classmate who had one recently? If so, you have observed an effect and inferred a cause. The event is the effect, and the reason for the event is the cause.

When scientists are unsure of the cause of a certain event, they design controlled experiments to determine what caused it.

Interpreting Data

The word *interpret* means "to explain the meaning of something." Look at the problem originally being explored in an experiment and figure out what the data show. Identify the control group and the test group so you can see whether or not changes in the independent variable have had an effect. Look for differences in the dependent variable between the control and test groups.

These differences you observe can be qualitative or quantitative. You would be able to describe a qualitative difference using only words, whereas you would measure a quantitative difference and describe it using numbers. If there are qualitative or quantitative differences, the independent variable that is being tested could have had an effect. If no qualitative or quantitative differences are found between the control and test groups, the variable that is being tested apparently had no effect.

For example, suppose that three pepper plants are placed in a garden and two of the plants are fertilized, but the third is left to grow without fertilizer. Suppose you are then asked to describe any differences in the plants after two weeks. A qualitative difference might be the appearance of brighter green leaves on fertilized plants but not on the unfertilized plant. A quantitative difference might be a difference in the height of the plants or the number of flowers on them.

Inferring Scientists often make inferences based on their observations. An inference is an attempt to explain, or interpret, observations or to indicate what caused what you observed. An inference is a type of conclusion.

When making an inference, be certain to use accurate data and accurately described observations. Analyze all of the data that you've collected. Then, based on everything you know, explain or interpret what you've observed.

Drawing Conclusions

When scientists have analyzed the data they collected, they proceed to draw conclusions about what the data mean. These conclusions are sometimes stated using words similar to those found in the hypothesis formed earlier in the process.

Conclusions To analyze your data, you must review all of the observations and measurements that you made and recorded. Recheck all data for accuracy. After your data are rechecked and organized, you are almost ready to draw a conclusion such as "Plants need sunlight in order to grow."

Before you can draw a conclusion, however, you must determine whether the data allow you to come to a conclusion that supports a hypothesis. Sometimes that will be the case, other times it will not.

If your data do not support a hypothesis, it does not mean that the hypothesis is wrong. It means only that the results of the investigation did not support the hypothesis. Maybe the experiment needs to be redesigned, but very likely, some of the initial observations on which the hypothesis was based were incomplete or biased. Perhaps more observation or research is needed to refine the hypothesis.

Avoiding Bias Sometimes drawing a conclusion involves making judgments. When you make a judgment, you form an opinion about what your data mean. It is important to be honest and to avoid reaching a conclusion if there were no supporting evidence for it or if it were based on a small sample. It also is important not to allow any expectations of results to bias your judgments. If possible, it is a good idea to collect additional data. Scientists do this all the time.

For example, animal behaviorist Katharine Payne made an important observation about elephant communication. While visiting a zoo, Payne felt the air vibrating around her. At the same time, she also noticed that the skin on an elephant's forehead was fluttering. She suspected that the elephants were generating the vibrations and that they might be using the low-frequency sounds to communicate.

Payne conducted an experiment to record these sounds and simultaneously observe the behavior of the elephants in the zoo. She later conducted a similar experiment in Namibia in southwest Africa, where elephant herds roam. The additional data she collected supported the judgment Payne had made, which was that these low-frequency sounds were a form of communication between elephants.

Evaluating Others' Data and Conclusions

Sometimes scientists have to use data that they did not collect themselves, or they have to rely on observations and conclusions drawn by other researchers. In cases such as these, the data must be evaluated carefully.

How were the data obtained? How was the investigation done? Has it been duplicated by other researchers? Did they come up with the same results? Look at the conclusion, as well. Would you reach the same conclusion from these results? Only when you have confidence in the data of others can you believe it is true and feel comfortable using it.

Communicating

The communication of ideas is an important part of the work of scientists. A discovery that is not reported will not advance the scientific community's understanding or knowledge. Communication among scientists also is important as a way of improving their investigations.

Scientists communicate in many ways, from writing articles in journals and magazines that explain their investigations and experiments, to announcing important discoveries on television and radio, to sharing ideas with colleagues on the Internet or presenting them as lectures.

Technology Skill Handbook

Computer Skills

People who study science rely on computers to record and store data and to analyze results from investigations. Whether you work in a laboratory or just need to write a lab report with tables, good computer skills are a necessity.

Using a Word Processor

Suppose your teacher has assigned a written report. After you've completed your research and decided how you want to write the information, you need to put all that information on paper. The easiest way to do this is with a word processing application on a computer.

A computer application that allows you to type your information, change it as many times as you need to, and then print it out so that it looks neat and clean is called a word processing application. You also can use this type of application to create tables and columns, add bullets or cartoon art to your page, include page numbers, and even check your spelling.

Helpful Hints

- If you aren't sure how to do something using your word processing program, look in the help menu. You will find a list of topics there to click on for help. After you locate the help topic you need, just follow the step-by-step instructions you see on your screen.
- Just because you've spell checked your report doesn't mean that the spelling is perfect. The spell check feature can't catch misspelled words that look like other words. If you've accidentally typed *wind* instead of *wing*, the spell checker won't know the difference. Always reread your report to make sure you didn't miss any mistakes.

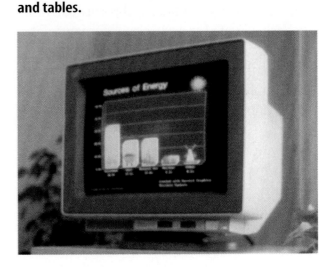

Figure 18
You can use computer programs to make graphs and tables.

Using a Database

Imagine you're in the middle of a research project busily gathering facts and information. You soon realize that it's becoming more difficult to organize and keep track of all the information. The tool to use to solve information overload is a database. Just as a file cabinet organizes paper records, a database organizes computer records. However, a database is more powerful than a simple file cabinet because at the click of a mouse, the contents can be reshuffled and reorganized. At computer-quick speeds, databases can sort information by any characteristics and filter data into multiple categories.

Helpful Hints

- Before setting up a database, take some time to learn the features of your database software by practicing with established database software.
- Periodically save your database as you enter data. That way, if something happens such as your computer malfunctions or the power goes off, you won't lose all of your work.

Technology Skill Handbook

Doing a Database Search

When searching for information in a database, use the following search strategies to get the best results. These are the same search methods used for searching Internet databases.

- Place the word *and* between two words in your search if you want the database to look for any entries that have both the words. For example, "fox *and* mink" would give you information that mentions both fox and mink.
- Place the word *or* between two words if you want the database to show entries that have at least one of the words. For example "fox *or* mink" would show you information that mentions either fox or mink.
- Place the word *not* between two words if you want the database to look for entries that have the first word but do not have the second word. For example, "canine *not* fox" would show you information that mentions the term canine but does not mention the fox.

In summary, databases can be used to store large amounts of information about a particular subject. Databases allow biologists, Earth scientists, and physical scientists to search for information quickly and accurately.

Using an Electronic Spreadsheet

Your science fair experiment has produced lots of numbers. How do you keep track of all the data, and how can you easily work out all the calculations needed? You can use a computer program called a spreadsheet to record data that involve numbers. A spreadsheet is an electronic mathematical worksheet.

Type in your data in rows and columns, just as in a data table on a sheet of paper. A spreadsheet uses simple math to do data calculations. For example, you could add, subtract, divide, or multiply any of the values in the spreadsheet by another number. You also could set up a series of math steps you want to apply to the data. If you want to add 12 to all the numbers and then multiply all the numbers by 10, the computer does all the calculations for you in the spreadsheet. Below is an example of a spreadsheet that records data from an experiment with mice in a maze.

Helpful Hints

- Before you set up the spreadsheet, identify how you want to organize the data. Include any formulas you will need to use.
- Make sure you have entered the correct data into the correct rows and columns.
- You also can display your results in a graph. Pick the style of graph that best represents the data with which you are working.

Figure 19
A spreadsheet allows you to display large amounts of data and do calculations automatically.

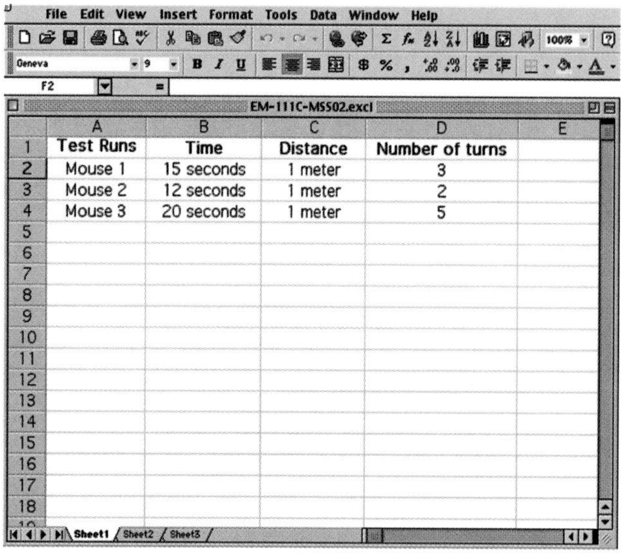

	A	B	C	D
1	Test Runs	Time	Distance	Number of turns
2	Mouse 1	15 seconds	1 meter	3
3	Mouse 2	12 seconds	1 meter	2
4	Mouse 3	20 seconds	1 meter	5

Technology Skill Handbook

Using a Computerized Card Catalog

When you have a report or paper to research, you probably go to the library. To find the information you need in the library, you might have to use a computerized card catalog. This type of card catalog allows you to search for information by subject, by title, or by author. The computer then will display all the holdings the library has on the subject, title, or author requested.

A library's holdings can include books, magazines, databases, videos, and audio materials. When you have chosen something from this list, the computer will show whether an item is available and where in the library to find it.

Helpful Hints

- Remember that you can use the computer to search by subject, author, or title. If you know a book's author but not the title, you can search for all the books the library has by that author.
- When searching by subject, it's often most helpful to narrow your search by using specific search terms, such as *and*, *or*, and *not*. If you don't find enough sources, you can broaden your search.
- Pay attention to the type of materials found in your search. If you need a book, you can eliminate any videos or other resources that come up in your search.
- Knowing how your library is arranged can save you a lot of time. The librarian will show you where certain types of materials are kept and how to find specific holdings.

Using Graphics Software

Are you having trouble finding that exact piece of art you're looking for? Do you have a picture in your mind of what you want but can't seem to find the right graphic to represent your ideas? To solve these problems, you can use graphics software. Graphics software allows you to create and change images and diagrams in almost unlimited ways. Typical uses for graphics software include arranging clip art, changing scanned images, and constructing pictures from scratch. Most graphics software applications work in similar ways. They use the same basic tools and functions. Once you master one graphics application, you can use any other graphics application relatively easily.

Figure 20
Graphics software can use your data to draw bar graphs.

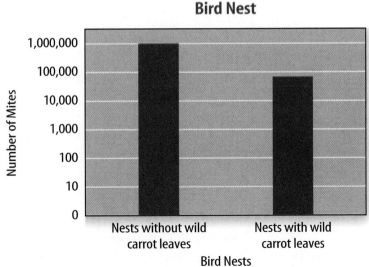

Technology Skill Handbook

Figure 21
Graphics software can use your data to draw circle graphs.

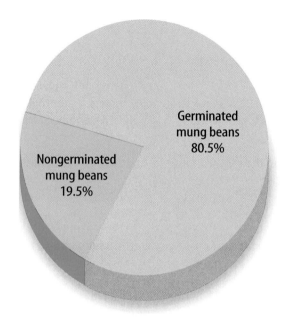

Helpful Hints
- As with any method of drawing, the more you practice using the graphics software, the better your results will be.
- Start by using the software to manipulate existing drawings. Once you master this, making your own illustrations will be easier.
- Clip art is available on CD-ROMs and the Internet. With these resources, finding a piece of clip art to suit your purposes is simple.
- As you work on a drawing, save it often.

Developing Multimedia Presentations

It's your turn—you have to present your science report to the entire class. How do you do it? You can use many different sources of information to get the class excited about your presentation. Posters, videos, photographs, sound, computers, and the Internet can help show your ideas.

First, determine what important points you want to make in your presentation. Then, write an outline of what materials and types of media would best illustrate those points. Maybe you could start with an outline on an overhead projector, then show a video, followed by something from the Internet or a slide show accompanied by music or recorded voices. You might choose to use a presentation builder computer application that can combine all these elements into one presentation. Make sure the presentation is well constructed to make the most impact on the audience.

Figure 22
Multimedia presentations use many types of print and electronic materials.

Helpful Hints
- Carefully consider what media will best communicate the point you are trying to make.
- Make sure you know how to use any equipment you will be using in your presentation.
- Practice the presentation several times.
- If possible, set up all of the equipment ahead of time. Make sure everything is working correctly.

Math Skill Handbook

Use this Math Skill Handbook to help solve problems you are given in this text. You might find it useful to review topics in this Math Skill Handbook first.

Converting Units

In science, quantities such as length, mass, and time sometimes are measured using different units. Suppose you want to know how many miles are in 12.7 km?

Conversion factors are used to change from one unit of measure to another. A conversion factor is a ratio that is equal to one. For example, there are 1,000 mL in 1 L, so 1,000 mL equals 1 L, or:

$$1{,}000 \text{ mL} = 1 \text{ L}$$

If both sides are divided by 1 L, this equation becomes:

$$\frac{1{,}000 \text{ mL}}{1 \text{ L}} = 1$$

The **ratio** on the left side of this equation is equal to one and is a conversion factor. You can make another conversion factor by dividing both sides of the top equation by 1,000 mL:

$$1 = \frac{1 \text{ L}}{1{,}000 \text{ mL}}$$

To **convert units,** you multiply by the appropriate conversion factor. For example, how many milliliters are in 1.255 L? To convert 1.255 L to milliliters, multiply 1.255 L by a conversion factor.

Use the **conversion factor** with new units (mL) in the numerator and the old units (L) in the denominator.

$$1.255 \text{ L} \times \frac{1{,}000 \text{ mL}}{1 \text{ L}} = 1{,}255 \text{ mL}$$

The unit L divides in this equation, just as if it were a number.

Example 1 There are 2.54 cm in 1 inch. If a meterstick has a length of 100 cm, how long is the meterstick in inches?

Step 1 Decide which conversion factor to use. You know the length of the meterstick in centimeters, so centimeters are the old units. You want to find the length in inches, so inch is the new unit.

Step 2 Form the conversion factor. Start with the relationship between the old and new units.

$$2.54 \text{ cm} = 1 \text{ inch}$$

Step 3 Form the conversion factor with the old unit (centimeter) on the bottom by dividing both sides by 2.54 cm.

$$1 = \frac{2.54 \text{ cm}}{2.54 \text{ cm}} = \frac{1 \text{ inch}}{2.54 \text{ cm}}$$

Step 4 Multiply the old measurement by the conversion factor.

$$100 \text{ cm} \times \frac{1 \text{ inch}}{2.54 \text{ cm}} = 39.37 \text{ inches}$$

The meter stick is 39.37 inches long.

Example 2 There are 365 days in one year. If a person is 14 years old, what is his or her age in days? (Ignore leap years)

Step 1 Decide which conversion factor to use. You want to convert years to days.

Step 2 Form the conversion factor. Start with the relation between the old and new units.

$$1 \text{ year} = 365 \text{ days}$$

Step 3 Form the conversion factor with the old unit (year) on the bottom by dividing both sides by 1 year.

$$1 = \frac{1 \text{ year}}{1 \text{ year}} = \frac{365 \text{ days}}{1 \text{ year}}$$

Step 4 Multiply the old measurement by the conversion factor:

$$14 \text{ years} \times \frac{365 \text{ days}}{1 \text{ year}} = 5{,}110 \text{ days}$$

The person's age is 5,110 days.

Practice Problem A cat has a mass of 2.31 kg. If there are 1,000 g in 1 kg, what is the mass of the cat in grams?

Math Skill Handbook

Using Fractions

A **fraction** is a number that compares a part to the whole. For example, in the fraction $\frac{2}{3}$, the 2 represents the part and the 3 represents the whole. In the fraction $\frac{2}{3}$, the top number, 2, is called the numerator. The bottom number, 3, is called the denominator.

Sometimes fractions are not written in their simplest form. To determine a fraction's **simplest form,** you must find the greatest common factor (GCF) of the numerator and denominator. The greatest common factor is the largest common factor of all the factors the two numbers have in common.

For example, because the number 3 divides into 12 and 30 evenly, it is a common factor of 12 and 30. However, because the number 6 is the largest number that evenly divides into 12 and 30, it is the **greatest common factor.**

After you find the greatest common factor, you can write a fraction in its simplest form. Divide both the numerator and the denominator by the greatest common factor. The number that results is the fraction in its **simplest form.**

Example Twelve of the 20 corn plants in a field are over 1.5 m tall. What fraction of the corn plants in the field are over 1.5 m tall?

Step 1 Write the fraction.

$$\frac{\text{part}}{\text{whole}} = \frac{12}{20}$$

Step 2 To find the GCF of the numerator and denominator, list all of the factors of each number.

Factors of 12: 1, 2, 3, 4, 6, 12 (the numbers that divide evenly into 12)

Factors of 20: 1, 2, 4, 5, 10, 20 (the numbers that divide evenly into 20)

Step 3 List the common factors.

1, 2, 4.

Step 4 Choose the greatest factor in the list of common factors.

The GCF of 12 and 20 is 4.

Step 5 Divide the numerator and denominator by the GCF.

$$\frac{12 \div 4}{20 \div 4} = \frac{3}{5}$$

In the field, $\frac{3}{5}$ of the corn plants are over 1.5 m tall.

Practice Problem There are 90 duck eggs in a population. Of those eggs, 66 hatch over a one-week period. What fraction of the eggs hatch over a one-week period? Write the fraction in simplest form.

Math Skill Handbook

Calculating Ratios

A **ratio** is a comparison of two numbers by division.

Ratios can be written 3 to 5 or 3:5. Ratios also can be written as fractions, such as $\frac{3}{5}$. Ratios, like fractions, can be written in simplest form. Recall that a fraction is in **simplest form** when the greatest common factor (GCF) of the numerator and denominator is 1.

Example From a package of sunflower seeds, 40 seeds germinated and 64 did not. What is the ratio of germinated to not germinated seeds as a fraction in simplest form?

Step 1 Write the ratio as a fraction.

$$\frac{\text{germinated}}{\text{not germinated}} = \frac{40}{64}$$

Step 2 Express the fraction in simplest form. The GCF of 40 and 64 is 8.

$$\frac{40}{64} = \frac{40 \div 8}{64 \div 8} = \frac{5}{8}$$

The ratio of germinated to not germinated seeds is $\frac{5}{8}$.

Practice Problem Two children measure 100 cm and 144 cm in height. What is the ratio of their heights in simplest fraction form?

Using Decimals

A **decimal** is a fraction with a denominator of 10, 100, 1,000, or another power of 10. For example, 0.854 is the same as the fraction $\frac{854}{1,000}$.

In a decimal, the decimal point separates the ones place and the tenths place. For example, 0.27 means twenty-seven hundredths, or $\frac{27}{100}$, where 27 is the **number of units** out of 100 units. Any fraction can be written as a decimal using division.

Example Write $\frac{5}{8}$ as a decimal.

Step 1 Write a division problem with the numerator, 5, as the dividend and the denominator, 8, as the divisor. Write 5 as 5.000.

Step 2 Solve the problem.

```
      0.625
   8)5.000
     48
     ---
      20
      16
      --
       40
       40
       --
        0
```

Therefore, $\frac{5}{8} = 0.625$.

Practice Problem Write $\frac{19}{25}$ as a decimal.

Math Skill Handbook

Using Percentages

The word *percent* means "out of one hundred." A **percent** is a ratio that compares a number to 100. Suppose you read that 77 percent of all fish on Earth live in the Pacific Ocean. That is the same as reading that the Earth's fish that live in the Pacific Ocean is $\frac{77}{100}$. To express a fraction as a percent, first find an equivalent decimal for the fraction. Then, multiply the decimal by 100 and add the percent symbol. For example, $\frac{1}{2} = 1 \div 2 = 0.5$. Then $0.5 = 0.50 = 50\%$.

Example Express $\frac{13}{20}$ as a percent.

Step 1 Find the equivalent decimal for the fraction.

$$\begin{array}{r} 0.65 \\ 20\overline{)13.00} \\ \underline{120} \\ 100 \\ \underline{100} \\ 0 \end{array}$$

Step 2 Rewrite the fraction $\frac{13}{20}$ as 0.65.

Step 3 Multiply 0.65 by 100 and add the % sign.

$0.65 \cdot 100 = 65 = 65\%$

So, $\frac{13}{20} = 65\%$.

Practice Problem In an experimental population of 365 sheep, 73 were brown. What percent of the sheep were brown?

Using Precision and Significant Digits

When you make a **measurement,** the value you record depends on the precision of the measuring instrument. When adding or subtracting numbers with different precision, the answer is rounded to the smallest number of decimal places of any number in the sum or difference. When multiplying or dividing, the answer is rounded to the smallest number of significant figures of any number being multiplied or divided. When counting the number of **significant figures,** all digits are counted except zeros at the end of a number with no decimal such as 2,500, and zeros at the beginning of a decimal such as 0.03020.

Example The lengths 5.28 and 5.2 are measured in meters. Find the sum of these lengths and report the sum using the least precise measurement.

Step 1 Find the sum.

5.28 m	2 digits after the decimal
+ 5.2 m	1 digit after the decimal
10.48 m	

Step 2 Round to one digit after the decimal because the least number of digits after the decimal of the numbers being added is 1.

The sum is 10.5 m.

Practice Problem Multiply the numbers in the example using the rule for multiplying and dividing. Report the answer with the correct number of significant figures.

Math Skill Handbook

Solving One-Step Equations

An **equation** is a statement that two things are equal. For example, $A = B$ is an equation that states that A is equal to B.

Sometimes one side of the equation will contain a **variable** whose value is not known. In the equation $3x = 12$, the variable is x.

The equation is solved when the variable is replaced with a value that makes both sides of the equation equal to each other. For example, the solution of the equation $3x = 12$ is $x = 4$. If the x is replaced with 4, then the equation becomes $3 \cdot 4 = 12$, or $12 = 12$.

To solve an equation such as $8x = 40$, divide both sides of the equation by the number that multiplies the variable.

$$8x = 40$$
$$\frac{8x}{8} = \frac{40}{8}$$
$$x = 5$$

You can check your answer by replacing the variable with your solution and seeing if both sides of the equation are the same.

$$8x = 8 \cdot 5 = 40$$

The left and right sides of the equation are the same, so $x = 5$ is the solution.

Sometimes an equation is written in this way: $a = bc$. This also is called a **formula**. The letters can be replaced by numbers, but the numbers must still make both sides of the equation the same.

Example 1 Solve the equation $10x = 35$.

Step 1 Find the solution by dividing each side of the equation by 10.

$$10x = 35 \qquad \frac{10x}{10} = \frac{35}{10} \qquad x = 3.5$$

Step 2 Check the solution.

$$10x = 35 \qquad 10 \times 3.5 = 35 \qquad 35 = 35$$

Both sides of the equation are equal, so $x = 3.5$ is the solution to the equation.

Example 2 In the formula $a = bc$, find the value of c if $a = 20$ and $b = 2$.

Step 1 Rearrange the formula so the unknown value is by itself on one side of the equation by dividing both sides by b.

$$a = bc$$
$$\frac{a}{b} = \frac{bc}{b}$$
$$\frac{a}{b} = c$$

Step 2 Replace the variables a and b with the values that are given.

$$\frac{a}{b} = c$$
$$\frac{20}{2} = c$$
$$10 = c$$

Step 3 Check the solution.

$$a = bc$$
$$20 = 2 \times 10$$
$$20 = 20$$

Both sides of the equation are equal, so $c = 10$ is the solution when $a = 20$ and $b = 2$.

Practice Problem In the formula $h = gd$, find the value of d if $g = 12.3$ and $h = 17.4$.

Math Skill Handbook

Using Proportions

A **proportion** is an equation that shows that two ratios are equivalent. The ratios $\frac{2}{4}$ and $\frac{5}{10}$ are equivalent, so they can be written as $\frac{2}{4} = \frac{5}{10}$. This equation is an example of a proportion.

When two ratios form a proportion, the **cross products** are equal. To find the cross products in the proportion $\frac{2}{4} = \frac{5}{10}$, multiply the 2 and the 10, and the 4 and the 5. Therefore $2 \cdot 10 = 4 \cdot 5$, or $20 = 20$.

Because you know that both proportions are equal, you can use cross products to find a missing term in a proportion. This is known as **solving the proportion.** Solving a proportion is similar to solving an equation.

Example The heights of a tree and a pole are proportional to the lengths of their shadows. The tree casts a shadow of 24 m at the same time that a 6-m pole casts a shadow of 4 m. What is the height of the tree?

Step 1 Write a proportion.

$$\frac{\text{height of tree}}{\text{height of pole}} = \frac{\text{length of tree's shadow}}{\text{length of pole's shadow}}$$

Step 2 Substitute the known values into the proportion. Let h represent the unknown value, the height of the tree.

$$\frac{h}{6} = \frac{24}{4}$$

Step 3 Find the cross products.

$$h \cdot 4 = 6 \cdot 24$$

Step 4 Simplify the equation.

$$4h = 144$$

Step 5 Divide each side by 4.

$$\frac{4h}{4} = \frac{144}{4}$$

$$h = 36$$

The height of the tree is 36 m.

Practice Problem The proportions of bluefish are stable by the time they reach a length of 30 cm. The distance from the tip of the mouth to the back edge of the gill cover in a 35-cm bluefish is 15 cm. What is the distance from the tip of the mouth to the back edge of the gill cover in a 59-cm bluefish?

Math Skill Handbook

Using Statistics

Statistics is the branch of mathematics that deals with collecting, analyzing, and presenting data. In statistics, there are three common ways to summarize the data with a single number—the mean, the median, and the mode.

The **mean** of a set of data is the arithmetic average. It is found by adding the numbers in the data set and dividing by the number of items in the set.

The **median** is the middle number in a set of data when the data are arranged in numerical order. If there were an even number of data points, the median would be the mean of the two middle numbers.

The **mode** of a set of data is the number or item that appears most often.

Another number that often is used to describe a set of data is the range. The **range** is the difference between the largest number and the smallest number in a set of data.

A **frequency table** shows how many times each piece of data occurs, usually in a survey. The frequency table below shows the results of a student survey on favorite color.

Color	Tally	Frequency
red	IIII	4
blue	IIII	5
black	II	2
green	III	3
purple	IIII II	7
yellow	IIII I	6

Based on the frequency table data, which color is the favorite?

Example The high temperatures (in °C) on five consecutive days in a desert habitat under study are 39°, 37°, 44°, 36°, and 44°. Find the mean, median, mode, and range of this set.

To find the mean:
Step 1 Find the sum of the numbers.
$$39 + 37 + 44 + 36 + 44 = 200$$
Step 2 Divide the sum by the number of items, which is 5.
$$200 \div 5 = 40$$

The mean high temperature is 40°C.

To find the median:
Step 1 Arrange the temperatures from least to greatest.
$$36, 37, \underline{39}, 44, 44$$
Step 2 Determine the middle temperature.

The median high temperature is 39°C.

To find the mode:
Step 1 Group the numbers that are the same together.
$$44, 44, 36, 37, 39$$
Step 2 Determine the number that occurs most in the set.
$$\underline{44, 44}, 36, 37, 39$$

The mode measure is 44°C.

To find the range:
Step 1 Arrange the temperatures from largest to smallest.
$$44, 44, 39, 37, 36$$
Step 2 Determine the largest and smallest temperature in the set.
$$\underline{44}, 44, 39, 37, \underline{36}$$
Step 3 Find the difference between the largest and smallest temperatures.
$$44 - 36 = 8$$

The range is 8°C.

Practice Problem Find the mean, median, mode, and range for the data set 8, 4, 12, 8, 11, 14, 16.

Reference Handbook A

Safety in the Science Classroom

1. Always obtain your teacher's permission to begin an investigation.

2. Study the procedure. If you have questions, ask your teacher. Be sure you understand any safety symbols shown on the page.

3. Use the safety equipment provided for you. Goggles and a safety apron should be worn during most investigations.

4. Always slant test tubes away from yourself and others when heating them or adding substances to them.

5. Never eat or drink in the lab, and never use lab glassware as food or drink containers. Never inhale chemicals. Do not taste any substances or draw any material into a tube with your mouth.

6. Report any spill, accident, or injury, no matter how small, immediately to your teacher, then follow his or her instructions.

7. Know the location and proper use of the fire extinguisher, safety shower, fire blanket, first aid kit, and fire alarm.

8. Keep all materials away from open flames. Tie back long hair and tie down loose clothing.

9. If your clothing should catch fire, smother it with the fire blanket, or get under a safety shower. NEVER RUN.

10. If a fire should occur, turn off the gas then leave the room according to established procedures.

Follow these procedures as you clean up your work area

1. Turn off the water and gas. Disconnect electrical devices.

2. Clean all pieces of equipment and return all materials to their proper places.

3. Dispose of chemicals and other materials as directed by your teacher. Place broken glass and solid substances in the proper containers. Make sure never to discard materials in the sink.

4. Clean your work area. Wash your hands thoroughly after working in the laboratory.

First Aid	
Injury	**Safe Response** ALWAYS NOTIFY YOUR TEACHER IMMEDIATELY
Burns	Apply cold water.
Cuts and Bruises	Stop any bleeding by applying direct pressure. Cover cuts with a clean dressing. Apply ice packs or cold compresses to bruises.
Fainting	Leave the person lying down. Loosen any tight clothing and keep crowds away.
Foreign Matter in Eye	Flush with plenty of water. Use eyewash bottle or fountain.
Poisoning	Note the suspected poisoning agent.
Any Spills on Skin	Flush with large amounts of water or use safety shower.

Reference Handbook B

SI—Metric/English, English/Metric Conversions

	When you want to convert:	To:	Multiply by:
Length	inches	centimeters	2.54
	centimeters	inches	0.39
	yards	meters	0.91
	meters	yards	1.09
	miles	kilometers	1.61
	kilometers	miles	0.62
Mass and Weight*	ounces	grams	28.35
	grams	ounces	0.04
	pounds	kilograms	0.45
	kilograms	pounds	2.2
	tons (short)	tonnes (metric tons)	0.91
	tonnes (metric tons)	tons (short)	1.10
	pounds	newtons	4.45
	newtons	pounds	0.22
Volume	cubic inches	cubic centimeters	16.39
	cubic centimeters	cubic inches	0.06
	liters	quarts	1.06
	quarts	liters	0.95
	gallons	liters	3.78
Area	square inches	square centimeters	6.45
	square centimeters	square inches	0.16
	square yards	square meters	0.83
	square meters	square yards	1.19
	square miles	square kilometers	2.59
	square kilometers	square miles	0.39
	hectares	acres	2.47
	acres	hectares	0.40
Temperature	To convert °Celsius to °Fahrenheit		°C × 9/5 + 32
	To convert °Fahrenheit to °Celsius		5/9 (°F − 32)

*Weight is measured in standard Earth gravity.

Reference Handbook C

Care and Use of a Microscope

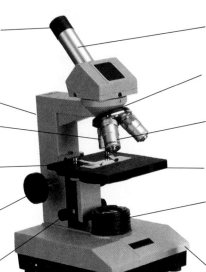

Eyepiece Contains magnifying lenses you look through.

Arm Supports the body tube.

Low-power objective Contains the lens with the lowest power magnification.

Stage clips Hold the microscope slide in place.

Fine adjustment Sharpens the image under high magnification.

Coarse adjustment Focuses the image under low power.

Body tube Connects the eyepiece to the revolving nosepiece.

Revolving nosepiece Holds and turns the objectives into viewing position.

High-power objective Contains the lens with the highest magnification.

Stage Supports the microscope slide.

Light source Provides light that passes upward through the diaphragm, the specimen, and the lenses.

Base Provides support for the microscope.

Caring for a Microscope

1. Always carry the microscope holding the arm with one hand and supporting the base with the other hand.

2. Don't touch the lenses with your fingers.

3. The coarse adjustment knob is used only when looking through the lowest-power objective lens. The fine adjustment knob is used when the high-power objective is in place.

4. Cover the microscope when you store it.

Using a Microscope

1. Place the microscope on a flat surface that is clear of objects. The arm should be toward you.

2. Look through the eyepiece. Adjust the diaphragm so light comes through the opening in the stage.

3. Place a slide on the stage so the specimen is in the field of view. Hold it firmly in place by using the stage clips.

4. Always focus with the coarse adjustment and the low-power objective lens first. After the object is in focus on low power, turn the nosepiece until the high-power objective is in place. Use ONLY the fine adjustment to focus with the high-power objective lens.

Making a Wet-Mount Slide

1. Carefully place the item you want to look at in the center of a clean, glass slide. Make sure the sample is thin enough for light to pass through.

2. Use a dropper to place one or two drops of water on the sample.

3. Hold a clean coverslip by the edges and place it at one edge of the water. Slowly lower the coverslip onto the water until it lies flat.

4. If you have too much water or a lot of air bubbles, touch the edge of a paper towel to the edge of the coverslip to draw off extra water and draw out unwanted air.

Diversity of Life: Classification of Living Organisms

A six-kingdom system of classification of organisms is used today. Two kingdoms—Kingdom Archaebacteria and Kingdom Eubacteria—contain organisms that do not have a nucleus and that lack membrane-bound structures in the cytoplasm of their cells. The members of the other four kingdoms have a cell or cells that contain a nucleus and structures in the cytoplasm, some of which are surrounded by membranes. These kingdoms are Kingdom Protista, Kingdom Fungi, Kingdom Plantae, and Kingdom Animalia.

Kingdom Archaebacteria

one-celled; some absorb food from their surroundings; some are photosynthetic; some are chemosynthetic; many found in extremely harsh environments including salt ponds, hot springs, swamps, and deep-sea hydrothermal vents

Kingdom Eubacteria

one-celled; most absorb food from their surroundings; some are photosynthetic; some are chemosynthetic; many are parasites; many are round, spiral, or rod-shaped; some form colonies

Kingdom Protista

Phylum Euglenophyta one-celled; photosynthetic or take in food; most have one flagellum; euglenoids

Phylum Bacillariophyta one-celled; photosynthetic; have unique double shells made of silica; diatoms

Phylum Dinoflagellata one-celled; photosynthetic; contain red pigments; have two flagella; dinoflagellates

Phylum Chlorophyta one-celled, many-celled, or colonies; photosynthetic; contain chlorophyll; live on land, in freshwater, or salt water; green algae

Phylum Rhodophyta most are many-celled; photosynthetic; contain red pigments; most live in deep, saltwater environments; red algae

Phylum Phaeophyta most are many-celled; photosynthetic; contain brown pigments; most live in saltwater environments; brown algae

Phylum Rhizopoda one-celled; take in food; are free-living or parasitic; move by means of pseudopods; amoebas

Kingdom Eubacteria
Bacillus anthracis

Phylum Chlorophyta
Desmids

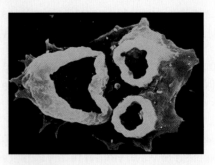

Amoeba

Phylum Zoomastigina one-celled; take in food; free-living or parasitic; have one or more flagella; zoomastigotes

Phylum Ciliophora one-celled; take in food; have large numbers of cilia; ciliates

Phylum Sporozoa one-celled; take in food; have no means of movement; are parasites in animals; sporozoans

Phylum Myxomycota
Slime mold

Phyla Myxomycota and Acrasiomycota one- or many-celled; absorb food; change form during life cycle; cellular and plasmodial slime molds

Phylum Oomycota many-celled; are either parasites or decomposers; live in freshwater or salt water; water molds, rusts and downy mildews

Kingdom Fungi

Phylum Zygomycota many-celled; absorb food; spores are produced in sporangia; zygote fungi; bread mold

Phylum Ascomycota one- and many-celled; absorb food; spores produced in asci; sac fungi; yeast

Phylum Basidiomycota many-celled; absorb food; spores produced in basidia; club fungi; mushrooms

Phylum Deuteromycota members with unknown reproductive structures; imperfect fungi; *Penicillium*

Mycophycota organisms formed by symbiotic relationship between an ascomycote or a basidiomycote and green alga or cyanobacterium; lichens

Phylum Oomycota
Phytophthora infestans

Lichens

Reference Handbook D

Kingdom Plantae

Divisions Bryophyta (mosses), **Anthocerophyta** (hornworts), **Hepatophytal** (liverworts), **Psilophytal** (whisk ferns) many-celled nonvascular plants; reproduce by spores produced in capsules; green; grow in moist, land environments

Division Lycophyta many-celled vascular plants; spores are produced in conelike structures; live on land; are photosynthetic; club mosses

Division Sphenophyta vascular plants; ribbed and jointed stems; scalelike leaves; spores produced in conelike structures; horsetails

Division Pterophyta vascular plants; leaves called fronds; spores produced in clusters of sporangia called sori; live on land or in water; ferns

Division Ginkgophyta deciduous trees; only one living species; have fan-shaped leaves with branching veins and fleshy cones with seeds; ginkgoes

Division Cycadophyta palmlike plants; have large, featherlike leaves; produces seeds in cones; cycads

Division Coniferophyta deciduous or evergreen; trees or shrubs; have needlelike or scalelike leaves; seeds produced in cones; conifers

Division Gnetophyta shrubs or woody vines; seeds are produced in cones; division contains only three genera; gnetum

Division Anthophyta dominant group of plants; flowering plants; have fruits with seeds

Kingdom Animalia

Phylum Porifera aquatic organisms that lack true tissues and organs; are asymmetrical and sessile; sponges

Phylum Cnidaria radially symmetrical organisms; have a digestive cavity with one opening; most have tentacles armed with stinging cells; live in aquatic environments singly or in colonies; includes jellyfish, corals, hydra, and sea anemones

Phylum Platyhelminthes bilaterally symmetrical worms; have flattened bodies; digestive system has one opening; parasitic and free-living species; flatworms

Division Anthophyta
Tomato plant

Division Bryophyta
Liverwort

Phylum Platyhelminthes
Flatworm

Reference Handbook D

Phylum Chordata

Phylum Nematoda round, bilaterally symmetrical body; have digestive system with two openings; free-living forms and parasitic forms; roundworms

Phylum Mollusca soft-bodied animals, many with a hard shell and soft foot or footlike appendage; a mantle covers the soft body; aquatic and terrestrial species; includes clams, snails, squid, and octopuses

Phylum Annelida bilaterally symmetrical worms; have round, segmented bodies; terrestrial and aquatic species; includes earthworms, leeches, and marine polychaetes

Phylum Arthropoda largest animal group; have hard exoskeletons, segmented bodies, and pairs of jointed appendages; land and aquatic species; includes insects, crustaceans, and spiders

Phylum Echinodermata marine organisms; have spiny or leathery skin and a water-vascular system with tube feet; are radially symmetrical; includes sea stars, sand dollars, and sea urchins

Phylum Chordata organisms with internal skeletons and specialized body systems; most have paired appendages; all at some time have a notochord, nerve cord, gill slits, and a postanal tail; include fish, amphibians, reptiles, birds, and mammals

English Glossary

This glossary defines each key term that appears in bold type in the text. It also shows the chapter, section, and page number where you can find the word used.

A

aggression: forceful behavior, such as fighting, used by an animal to control or dominate another animal in order to protect their young, defend territory, or get food. (Chap. 5, Sec. 2, p. 142)

amniotic egg: egg covered with a leathery shell that provides a complete environment for the embryo's development; for reptiles, a major adaptation for living on land. (Chap. 3, Sec. 4, p. 91)

anus: opening at the end of the digestive tract through which wastes leave the body. (Chap. 1, Sec. 3, p. 25)

appendages: jointed structures of arthropods, such as legs, wings, or antennae. (Chap. 2, Sec. 3, p. 48)

B

behavior: the way in which an organism interacts with other organisms and its environment; can be innate or learned. (Chap. 5, Sec. 1, p. 134)

bilateral symmetry: body parts arranged in a similar way on both sides of the body, with each half being a mirror image of the other half. (Chap. 1, Sec. 1, p. 13)

C

carnivore: animal that eats only other animals or the remains of other animals. (Chap. 1, Sec. 1, p. 9) (Chap. 4, Sec. 2, p. 115)

cartilage: tough, flexible tissue that joins vertebrae and makes up all or part of the vertebrate endoskeleton. (Chap. 3, Sec. 1, p. 73)

chordate: animal that has a notochord, a nerve cord, gill slits, and a postanal tail present at some stage in its development. (Chap. 3, Sec. 1, p. 72)

closed circulatory system: blood circulation system in which blood moves through the body in closed vessels. (Chap. 2, Sec. 1, p. 40)

conditioning: occurs when the response to a stimulus becomes associated with another stimulus. (Chap. 5, Sec. 1, p. 138)

contour feathers: strong, lightweight feathers that give birds their coloring and shape and that are used for flight. (Chap. 4, Sec. 1, p. 108)

courtship behavior: behavior that allows males and females of the same species to recognize each other and prepare to mate. (Chap. 5, Sec. 2, p. 143)

crop: digestive system sac in which earthworms store ingested soil. (Chap. 2, Sec. 2, p. 44)

cyclic behavior: behavior that occurs in repeated patterns. (Chap. 5, Sec. 2, p. 146)

D

down feathers: soft, fluffy feathers that provide an insulating layer next to the skin of adult birds and that cover the bodies of young birds. (Chap. 4, Sec. 1, p. 108)

English Glossary

E

ectotherm: vertebrate animal whose internal temperature changes when the temperature of its environment changes. (Chap. 3, Sec. 1, p. 75)

endoskeleton: supportive framework of bone and/or cartilage that provides an internal place for muscle attachment and protects a vertebrate's internal organs. (Chap. 3, Sec. 1, p. 73)

endotherm: vertebrate animal with a constant internal temperature. (Chap. 3, Sec. 1, p. 75)

estivation: inactivity in hot, dry months during which amphibians hide in cooler ground. (Chap. 3, Sec. 3, p. 85)

exoskeleton: thick, hard, outer covering that protects and supports arthropod bodies and provides places for muscles to attach. (Chap. 2, Sec. 3, p. 48)

F

fin: fanlike structure used by fish for steering, balancing, and movement. (Chap. 3, Sec. 2, p. 77)

free-living organism: organism that does not depend on another organism for food or a place to live. (Chap. 1, Sec. 3, p. 22)

G

gestation period: period during which an embryo develops in the uterus; the length of time varies among species. (Chap. 4, Sec. 2, p. 119)

gills: organs that exchange carbon dioxide for oxygen in the water. (Chap. 2, Sec. 1, p. 38)

gill slits: in developing chordates, the paired openings found in the area between the mouth and digestive tube. (Chap. 3, Sec. 1, p. 73)

gizzard: muscular digestive system structure in which earthworms grind soil and organic matter. (Chap. 2, Sec. 2, p. 44)

H

herbivore: animal that eats only plants or parts of plants. (Chap. 1, Sec. 1, p. 9) (Chap. 4, Sec. 2, p. 115)

hermaphrodite (hur MA fruh dite): animal that produces both sperm and eggs in the same body, but its sperm cannot fertilize its own eggs. (Chap. 1, Sec. 2, p. 16)

hibernation: cyclic response of inactivity and slowed metabolism that occurs during periods of cold temperatures and limited food supplies. (Chap. 3, Sec. 3, p. 85) (Chap. 5, Sec. 2, p. 147)

I

imprinting: occurs when an animal forms a social attachment to another organism during a specific period following birth or hatching. (Chap. 5, Sec. 1, p. 137)

incubate (IHN kyuh bayt): to keep eggs warm until they hatch; the length of time varies among species. (Chap. 4, Sec. 1, p. 106)

innate behavior: behavior that an organism is born with and does not have to be learned, such as a reflex or instinct. (Chap. 5, Sec. 1, p. 135)

English Glossary

insight: form of reasoning that allows animals to use past experiences to solve new problems. (Chap. 5, Sec. 1, p. 139)

instinct: complex pattern of innate behavior, such as spinning a web, that can take weeks to complete. (Chap. 5, Sec. 1, p. 136)

invertebrate: animal without a backbone. (Chap. 1, Sec. 1, p. 12)

M

mammals: endothermic vertebrates that have hair, teeth specialized for eating certain foods, and whose females have mammary glands that produce milk for feeding their young. (Chap. 4, Sec. 2, p. 114)

mammary glands: milk-producing glands of female mammals used to feed their young. (Chap. 4, Sec. 2, p. 114)

mantle: thin layer of tissue that covers a mollusk's body organs; secretes the shell or protects the body of mollusks without shells. (Chap. 2, Sec. 1, p. 38)

marsupial: a mammal with an external pouch for the development of its immature young. (Chap. 4, Sec. 2, p. 118)

medusa (mih DEW suh): cnidarian body type that is bell-shaped and free-swimming. (Chap. 1, Sec. 2, p. 17)

metamorphosis: process in which many insect species change their body form to become adults; can be complete (egg, larva, pupa, adult) or incomplete (egg, nymph, adult). (Chap. 2, Sec. 3, p. 50)

migration: instinctive seasonal movement of animals to find food or to reproduce in better conditions. (Chap. 5, Sec. 2, p. 148)

molting: shedding and replacing of an arthropod's exoskeleton. (Chap. 2, Sec. 3, p. 48)

monotreme: a mammal that lays eggs with tough, leathery shells and whose mammary glands do not have nipples. (Chap. 4, Sec. 2, p. 118)

N

nerve cord: tubelike structure above the notochord that in most chordates develops into the brain and spinal cord. (Chap. 3, Sec. 1, p. 73)

notochord: firm but flexible structure that extends along the upper part of a chordate's body. (Chap. 3, Sec. 1, p. 72)

O

omnivore: animal that eats plants and animals or animal flesh. (Chap. 1, Sec. 1, p. 9) (Chap. 4, Sec. 2, p. 114)

open circulatory system: blood circulation system in which blood moves through vessels and into open spaces around the body organs. (Chap. 2, Sec. 1, p. 38)

P

pheromone (FER uh mohn): powerful chemical produced by an animal to influence the behavior of another animal of the same species. (Chap. 5, Sec. 2, p. 143)

English Glossary

placenta: a saclike organ in which a placental embryo develops and that absorbs oxygen and food from the mother's blood. (Chap. 4, Sec. 2, p. 119)

placental: a mammal whose offspring develop inside a placenta in the female's uterus. (Chap. 4, Sec. 2, p. 119)

polyp (PAH lup): cnidarian body type that is vase-shaped and is usually sessile. (Chap. 1, Sec. 2, p. 17)

postanal tail: muscular structure at the end of a developing chordate. (Chap. 3, Sec. 1, p. 72)

preening: process in which a bird rubs oil from an oil gland over its feathers to condition them and make them water repellent. (Chap. 4, Sec. 1, p. 108)

R

radial symmetry: body parts arranged in a circle around a central point. (Chap. 1, Sec. 1, p. 13)

radula (RA juh luh): in gastropods, the tonguelike organ with rows of teeth used to scrape and tear food. (Chap. 2, Sec. 1, p. 39)

reflex: simple innate behavior, such as yawning or blinking, that is an automatic response and does not involve a message to the brain. (Chap. 5, Sec. 1, p. 135)

S

scales: thin, hard plates that cover a fish's skin and protect its body. (Chap. 3, Sec. 2, p. 77)

sessile (SE sile): describes an organism that remains attached to one place during its lifetime. (Chap. 1, Sec. 2, p. 15)

setae (SEE tee): bristlelike structures on the outside of each body segment that helps segmented worms move. (Chap. 2, Sec. 2, p. 43)

social behavior: interactions among members of the same species, including courtship and mating, getting food, caring for young, and protecting each other. (Chap. 5, Sec. 2, p. 140)

society: a group of animals of the same species that live and work together in an organized way, with each member doing a specific job. (Chap. 5, Sec. 2, p. 141)

spiracles (SPIHR ih kulz): openings in the abdomen and thorax of insects through which air enters and waste gases leave. (Chap. 2, Sec. 3, p. 49)

stinging cells: capsules with coiled triggerlike structures that help cnidarians capture food. (Chap. 1, Sec. 2, p. 18)

T

tentacles (TEN tih kulz): armlike structures that have stinging cells and surround the mouths of most cnidarians. (Chap. 1, Sec. 2, p. 18)

tube feet: hydraulic, hollow, thin-walled tubes that end in suction cups and enable echinoderms to move. (Chap. 2, Sec. 4, p. 58)

U

umbilical cord: connects the embryo to the placenta; moves food and oxygen from the placenta to the embryo and removes the embryo's waste products. (Chap. 4, Sec. 2, p. 119)

English Glossary

V

vertebrate: animal with a backbone. (Chap. 1, Sec. 1, p. 12) (Chap. 3, Sec. 1, p. 73)

W

water-vascular system: network of water-filled canals that allows echinoderms to move, capture food, give off wastes, and exchange carbon dioxide and oxygen. (Chap. 2, Sec. 4, p. 58)

Spanish Glossary

Este glosario define cada término clave que aparece en negrillas en el texto. También muestra el capítulo, la sección y el número de página en donde se usa dicho término.

A

aggression / agresión: comportamiento enérgico, como las peleas, que usa un animal para controlar o dominar a otro animal con el propósito de proteger sus crías, defender su territorio u obtener alimento. (Cap. 5, Sec. 2, pág. 142)

amniotic egg / huevo amniótico: huevo cubierto por una concha correosa que provee un ambiente completo para el desarrollo del embrión; para los reptiles es una adaptación importante para la vida en tierra. (Cap. 3, Sec. 4, pág. 91)

anus / ano: abertura al final del tracto digestivo por donde se expulsan los desechos del cuerpo. (Cap. 1, Sec. 3, pág. 25)

appendages / apéndices: estructuras adjuntas de los artrópodos, como piernas, alas o antenas. (Cap. 2, Sec. 3, pág. 48)

B

behavior / comportamiento: la interacción de un organismo con otro organismo y su medio ambiente; puede ser innato o adquirido. (Cap. 5, Sec. 1, pág. 134)

bilateral symmetry / simetría bilateral: partes corporales distribuidas de manera semejante en ambos lados del cuerpo, de tal forma que cada mitad es un reflejo exacto de la otra mitad. (Cap. 1, Sec. 1, pág. 13)

C

carnivore / carnívoro: animal que sólo se alimenta de otros animales o de los restos de otros animales. (Cap. 1, Sec. 1, pág. 9; Cap. 4 Sec. 2, pág. 115)

cartilage / cartílago: tejido flexible y duro que une las vertebras y que compone todo o parte del endoesqueleto vertebrado. (Cap. 3, Sec. 1, pág. 73)

chordate / cordado: animal que posee un notocordio, un cordón nervioso, hendiduras branquiales y una cola en alguna etapa de su desarrollo. (Cap. 3, Sec. 1, pág. 72)

closed circulatory system / sistema circulatorio cerrado: sistema de circulación en que la sangre se mueve a través del cuerpo en vasos cerrados. (Cap. 2, Sec. 1, pág. 40)

conditioning / condicionamiento: ocurre cuando la respuesta a un estímulo se asocia con otro estímulo. (Cap. 5, Sec. 1, pág. 138)

contour feathers / plumas de contorno: plumas livianas y fuertes que dan a las aves sus coloridos y formas y que se usan para el vuelo. (Cap. 4, Sec. 1, pág. 108)

courtship behavior / comportamiento de cortejo: comportamiento que permite que machos y hembras de una especie se reconozcan mutuamente y se preparen para el apareo. (Cap. 5, Sec. 2, pág. 143)

crop / buche: saco del sistema digestivo en el cual almacenan la tierra las lombrices de tierra. (Cap. 2, Sec. 2, pág. 44)

Spanish Glossary

cyclic behavior / comportamiento cíclico: comportamiento que ocurre en forma de patrones repetidos. (Cap. 5, Sec. 2, pág. 146)

D

down feathers / plumones: plumas esponjosas y suaves que proveen una capa de aislamiento cerca de la piel de las aves adultas y que cubren los cuerpos de las aves jóvenes. (Cap. 4, Sec. 1, pág. 108)

E

ectotherm / de sangre fría: animal vertebrado cuya temperatura interna cambia con los cambios en temperatura de su ambiente. (Cap. 3. Sec. 1. pág. 75)

endoskeleton / endoesqueleto: marco de apoyo para huesos y cartílago que provee un lugar interno para la adhesión de los músculos y protege los órganos internos de los vertebrados. (Cap. 3, Sec. 1. pg. 73)

endotherm / de sangre caliente: animal vertebrado con una temperatura corporal interna constante. (Cap, 3, Sec. 1, pág. 75)

estivation / estivación: inactividad en los meses calurosos y secos durante los cuales los anfibios se esconden dentro del suelo que está más fresco. (Cap. 3, Sec. 3, pág. 85)

exoskeleton / exoesqueleto: cubierta externa dura y gruesa que protege y da apoyo al cuerpo de los artrópodos, además de proveer un lugar para que se adhieran los músculos. (Cap. 2, Sec. 3, pág. 48)

F

fin / aleta: estructura en forma de abanico que usan los peces para cambiar de dirección, para el equilibrio y el movimiento. (Cap. 3, Sec. 2, pág. 77)

free-living organism / organismo de vida libre: organismo que no depende de otro organismo para obtener alimento o un lugar para vivir. (Cap. 1, Sec. 3, pág. 22)

G

gestation period / período de gestación: período en que un embrión se desarrolla en el útero; el período de gestación varía según la especie. (Cap. 4, Sec. 2, pág. 119)

gills / branquias: órganos que intercambian el dióxido de carbono por oxígeno en el agua. (Cap. 2, Sec. 1, pág. 38)

gill slits / hendiduras branquiales: en los cordados en etapa de desarrollo, los pares de aberturas que se encuentran en el área entre la boca y el tubo digestivo. (Cap. 3, Sec. 1, pág. 73)

gizzard / molleja: estructura muscular del sistema digestivo en la cual la lombriz de tierra muele la tierra y la materia orgánica. (Cap. 2, Sec. 2, pág. 44)

Spanish Glossary

H

herbivore / herbívoro: animal que se alimenta solamente de plantas o partes de plantas. (Cap. 1, Sec. 1, pág. 9; Cap. 4, Sec. 2, pág. 115)

hermaphrodite / hermafrodita: animal que produce tanto espermatozoides como huevos en el mismo cuerpo, pero cuyos espermatozoides no pueden fecundar sus propios huevos. (Cap. 1, Sec. 2, pág. 16)

hibernation / hibernación: respuesta cíclica de inactividad y disminución del metabolismo, la cual ocurre durante períodos de temperaturas frías y abastecimientos limitados de alimentos. (Cap. 3, Sec. 3, pág. 85; Cap. 5, Sec. 2, pág. 147)

I

imprinting / impronta: ocurre cuando un animal forma un vínculo social con otro organismo durante un período específico después del nacimiento o de salir del cascarón. (Cap. 5, Sec. 1, pág. 137)

incubate / incubar: mantener los huevos cálidos hasta que las crías salgan del cascarón; el período de incubación varía según la especie. (Cap. 4, Sec. 1, pág. 106)

innate behavior / comportamiento innato: comportamiento con que nace un organismo y el cual no tiene que ser adquirido, como un reflejo o un instinto. (Cap. 5, Sec. 1, pág. 135)

insight / discernimiento: forma de razonamiento que permite a los animales usar las experiencias previas para resolver nuevos problemas. (Cap. 5, Sec. 1, pág. 139)

instinct / instinto: patrón complejo de comportamiento innato, como por ejemplo, tejer una telaraña y el que puede demorar semanas en completarse. (Cap. 5, Sec. 1, pág. 136)

invertebrate / invertebrado: animal que carece de columna vertebral. (Cap. 1, Sec. 1, pág. 12)

M

mammals / mamíferos: vertebrados de sangre caliente que tienen pelo, dientes especializados para comer ciertos alimentos y cuyas hembras poseen glándulas mamarias que producen leche para alimentar a sus crías. (Cap. 4, Sec. 2, pág. 114)

mammary glands / glándulas mamarias: glándulas productoras de leche en las hembras de los mamíferos y las cuales usan para alimentar a sus crías. (Cap. 4, Sec. 2, pág. 114)

mantle / manto: capa delgada de tejido que cubre los órganos corporales de los moluscos; secreta la concha o protege el cuerpo de los moluscos que no tienen conchas. (Cap. 2, Sec. 1, pág. 38)

marsupial / marsupial: mamífero que posee una bolsa externa para el desarrollo de su cría inmadura. (Cap. 4, Sec. 2, pág. 118)

medusa / medusa: forma libre de diversos grupos de cnidarios que tiene forma acampanada. (Cap. 1, Sec. 2, pág. 17)

metamorphosis / metamorfosis: proceso mediante el cual muchas especies de insectos cambian su forma corporal para convertirse en adultos; puede ser completa (huevo, larva, crisálida, adulto) o incompleta (huevo, ninfa, adulto). (Cap. 2, Sec. 3, pág. 50)

Spanish Glossary

migration / migración: movimiento instintivo de ciertos animales de mudarse a lugares nuevos cuando cambian las estaciones, en busca de alimentos o para encontrar condiciones más propicias para el apareo. (Cap. 5, Sec. 2, pág. 148)

molting / muda: desprendimiento y reemplazo del exoesqueleto de un artrópodo. (Cap. 2, Sec. 3, pág. 48)

monotreme / monotrema: mamífero que pone huevos con cáscara dura y correosa y cuyas glándulas mamarias no tienen pezones. (Cap. 4, Sec. 2, pág. 118)

N

nerve cord / cordón nervioso: estructura tubular sobre el notocordio la cual se desarrolla en el encéfalo y la espina dorsal en la mayoría de los cordados. (Cap. 3, Sec. 1, pág. 73)

notochord / notocordio: estructura firme pero flexible que se extiende a lo largo de la parte superior del cuerpo de un cordado. (Cap. 3, Sec. 1, pág. 72)

O

omnivore / omnívoro: animal que se alimenta tanto de plantas como de animales. (Cap. 1, Sec. 1, pág. 9; Cap. 4, Sec. 2, pág. 114)

open circulatory system / sistema circulatorio abierto: sistema circulatorio en que la sangre se mueve a través de vasos y entre los espacios alrededor de los órganos corporales. (Cap. 2, Sec. 1, pág. 38)

P

pheromone / feromona: poderosa sustancia química producida por un animal para influir sobre el comportamiento de otro animal de la misma especie. (Cap. 5, Sec. 2, pág. 143)

placenta / placenta: órgano que parece un saco en el cual se desarrolla un embrión placentario y que absorbe oxígeno y alimento de la sangre de la madre. (Cap. 4, Sec. 2, pág. 119)

placental / placentario: mamífero cuyas crías se desarrollan dentro de una placenta en el útero de las hembras. (Cap. 4, Sec. 2, pág. 119)

polyp / pólipo: forma sésil de los cnidarios de forma tubular,. (Cap. 1, Sec. 2, pág. 17)

postanal tail / cola: estructura muscular en el extremo del cuerpo de un cordado en desarrollo. (Cap. 3, Sec. 1, pág. 72)

preening / arreglarse las plumas con el pico: proceso en que un ave se frota aceite de una glándula sebácea en sus plumas para arreglárselas y hacerlas resistentes al agua. (Cap. 4, Sec. 1, pág. 108)

R

radial symmetry / simetría radial: partes corporales distribuidas en círculo alrededor de un punto central. (Cap. 1, Sec. 1, pág. 13)

radula / rádula: en los gasterópodos, el órgano que parece una lengua con hileras de dientes y que usan para raspar y rasgar el alimento. (Cap. 2, Sec. 1, pág. 39)

reflex / reflejo: comportamiento innato simple, como bostezar o parpadear, que es una respuesta automática y que no

Spanish Glossary

involucra el envío de un mensaje al encéfalo. (Cap. 5, Sec. 1, pág. 135)

S

scales / escamas: placas duras y delgadas que cubren la piel de un pez y le protegen el cuerpo. (Cap. 3, Sec. 2, pág. 77)

sessile / sésil: se describe así un organismo que permanece adherido a un lugar durante toda su vida. (Cap. 1, Sec. 2, pág. 15)

setae / setas: estructuras que parecen cerdas en el exterior de cada segmento corporal y que facilita la locomoción de los gusanos. (Cap. 2, Sec. 2, pág. 43)

social behavior / comportamiento social: interacciones entre los miembros de la misma especie; incluye el comportamiento de cortejo, el apareo, la obtención de alimentos, el cuidado de las crías y la protección mutua. (Cap. 5, Sec. 2, pág. 140)

society / sociedad: grupo de animales de la misma especie que viven y trabajan juntos de manera organizada en la que cada cual realiza una tarea específica. (Cap. 5, Sec. 2, pág. 141)

spiracles / espiráculos: aberturas en el abdomen y tórax de insectos, a través de la cual entra el aire y salen los gases de desperdicio. (Cap. 2, Sec. 3, pág. 49)

stinging cells: / células urticantes o nematocistos: cápsulas con estructuras embobinadas presentes en los tentáculos de los cnidarios que se pueden lanzar a manera de arpón para ayudar a capturar el alimento. (Cap. 1, Sec. 2, pág. 18)

T

tentacles / tentáculos: estructuras móviles que poseen células urticantes y que rodean la boca de la mayoría de los cnidarios. (Cap. 1, Sec. 2, pág. 18)

tube feet / pedicelos: tubos hidráulicos de paredes huecas y delgadas que terminan en unas copas de succión y permiten la locomoción a los equinodermos. (Cap. 2, Sec. 4, pág. 58)

U

umbilical cord / cordón umbilical: conecta el embrión a la placenta; mueve alimentos y oxígeno de la placenta al embrión y elimina los productos de desecho del embrión. (Cap. 4, Sec. 2, pág. 119)

V

vertebrate / vertebrado: animal provisto de columna vertebral. (Cap. 1, Sec. 1, pág. 12; Cap. 3, Sec. 1, pág. 73)

W

water-vascular system / sistema vascular acuático: red de canales llenos de agua que permiten que los equinodermos se muevan, capturen alimentos, se deshagan de los desechos e intercambien dióxido de carbono y oxígeno. (Cap. 2, Sec. 4, pág. 58)

Index

The index for *Animal Diversity* will help you locate major topics in the book quickly and easily. Each entry in the index is followed by the number of the pages on which the entry is discussed. A page number given in boldfaced type indicates the page on which that entry is defined. A page number given in italic type indicates a page on which the entry is used in an illustration or photograph. The abbreviation *act.* indicates a page on which the entry is used in an activity.

A

Aardvark, 119
Activities, 21, 28–29, 57, 62–63, 76, 96–97, 123, 124–125, 149, 150–151
Adaptations, 9–11; behavioral, 11, *11*; for obtaining energy, 9, *9*; physical, 10, *10*; predator, 11, *11*
Aggression, 142
Agriculture: and arthropods, 55
Amniotic egg, 91, *91*
Amphibians, 74, *74*, 85–89; characteristics of, 85–86, *85*, *86*; frogs, 86, *86*, 87, *87*, 88, *88*, *89*; importance of, 88–89, *89*; metamorphosis of, 86–87, *86–87*; origin of, 89; salamanders, 85, *85*, 86, 88; toads, 87
Anglerfish, 83, *83*
Animal(s), 6–29. *see* Vertebrate animals. *see also* Invertebrate animals; adaptations of, 9–11, *9*, *10*, *11*; aggression in, 142; characteristics of, 8, *8*; classifying, 12–13, *12*, *13*; communication of, *act.* 133, 142–146, *142*, *144*, *145*; conditioning of, 138, *138*; courtship behavior of, 143, *143*; cyclic behavior of, 146–149, *146*, *147*, *148*, *act.* 149; habitats of, *act.* 150–151, *150*, *151*; hibernation of, 147, *147*; imprinting of, 137, *137*; innate behavior of, 134–135, *134*, *135*; instincts of, 136, *136*, 140; learned behavior of, 136–139, *136*, *137*, *138*, *139*; migration of, 148, *148*; plantlike, *8*; reflexes of, 135; social behavior of, 140–141, *140*, *141*; submission in, 142, *142*; territorial behavior of, 141, *141*
Annelids. *see* Segmented worms
Anus, 25
Appendage, 48
Appendices. *see* Reference Handbook
Arachnids, 52–53, *53*
Archaeopteryx, 75, 113, *113*
Archerfish, 79
Arthropods, 48–57; arachnids, 52–53, *53*; centipedes, 53, *53*, *54*; characteristics of, 48, *48*; crustaceans, 55, *55*, *act.* 57; diversity of, *54*; exoskeletons of, 48; insects, 49–51, *49*, *50*, *51*, *56*; millipedes, 53, *53*; origin of, 56, *56*; segmented bodies of, 48, *49*; value of, 55–56, *56*
Asexual reproduction, 16, 19, *19*, 23, *23*

B

Backbone, 12
Barnacles, 54, *55*
Bats, 120, *120*, 122; hibernation of, *147*
Beaks, *act.* 123
Bear, 115, *115*
Before You Read, 7, 37, 71, 105, 133
Behavior, 134, *134*; conditioned, 138, *138*; courtship, **143,** *143*; cyclic, 146–149, *146*, *147*, *148*, *act.* 149; innate, **134**–135, *134*, *135*; learned, **136**–139, *136*, *137*, *138*, *139*; of packs, 134, *134*, 141; social, **140**–141, *140*, *141*; territorial, 141, *141*
Behavioral adaptations, 11, *11*
Bilateral symmetry, 13, *13*
Biodiversity: of arthropods, 54
Bioluminescence, 144–146, *145*
Birds, 74, *74*, 104, *104*, 106–113, *124*, *125*; beaks of, *act.* 123; body systems of, 110, *110*; characteristics of, 106–109; courtship behavior of, 143, *143*; cyclic behavior of, 146, *146*; eggs of, 106, *106*, 126–127, *126*, *127*; feathers of, 108, *108*; flight adaptations of, 107–109, *107*, *108*, *109*; gizzards of, *act.* 105; hollow bones of, 107, *107*; importance of, 111, *111*; innate behavior of, 135, *135*; learned behavior of, 136, *136*, 137, *137*; origin of, 113, *113*; preening of, 108; reproduction of, 106, *106*; sound communication of, 144, *144*; types of, *112*; uses of, 111; watching, *act.* 124–125; wings of, 109, *109*
Bivalves, 40, *40*
Blubber, 116
Body systems: of birds, 110, *110*; of mammals, 116–117, *116*
Bones: of birds, 107, *107*
Bony fish, 74, *74*, 81–83, *81*, *83*
Book lungs, 53
Bristleworms, 45, *45*
Brittle stars, 59, *59*
Budding, 16
Butterflies, *51*, *54*, 148, *148*

Index

C

Camouflage, 10, 11, *11*, 51
Cardinalfish, 9, *9*
Carnivores, 9, *9*, **115**, *115*, 120, *120*, 122
Cartilage, 73, *73*
Cartilaginous fish, 74, *74*, 80
Castings: of earthworms, 44, *44*
Cell(s): in animals, 8; collar, 15, *15;* stinging, **18,** *18*
Centipedes, 53, *53*, 54
Cephalopods, 40–41, *40, 41*
Chameleons, 10, *11*
Champosaur, 94, *94*
Chemical communication, 143
Chemistry Integration, 16, 144
Chimpanzees, 139, *139*
Chitons, 9, *9*
Chordates, 72–73, *72, 73*
Cilia, 22
Circulatory system: of amphibians, 85; of birds, 110; closed, **40;** of earthworms, 44, *44;* of mammals, 117; open, **38**
Clams: *act.* 37, 40, 42
Classification: of animals, 12–13, *12, 13;* of mollusks, 39–41
Closed circulatory system, 40
Clown fish, 18
Cnidarians, 17–20, *17, 18, 19, 20, act.* 21
Cockroaches, 54
Coelacanths, 83
Collar cells, 15, *15*
Communication: of animals, *act.* 133, 142–146, *142, 144, 145;* chemical, 143; light, 144–146, *145;* sound, 144, *144*
Conches, 39, *39*
Conditioning, 138, *138*
Contour feathers, 108, *108*
Coral, 6, 17, 20
Coral reef, 20, *20*
Courtship behavior, 143, *143*
Crabs, 48, *48*, 54, 55, *55*
Crayfish, 55, *act.* 57
Crocodiles, 90, 93, *93*
Crop: of earthworms, **44,** *44*
Crustaceans, 55, *55, act.* 57
Cuttlefish, 40, *40*, 42
Cyclic behavior, 146–149, *146, 147, 148, act.* 149

D

Density: calculating, 82
Design Your Own Experiment, 28–29, 96–97,
Detritus, 9, 58, 60
Digestive system: of birds, 110; of earthworms, 44; of mammals, 116, *116;* microbes in, 9
Dinosaurs: extinction of, 95; fossils of, 75
Diseases: flatworms and, 23, *23, 26;* insects and, 51; roundworms and, 25, *26*, 27; ticks and, 53
Dogs: behavior of, 134, *134*, 138, *138*
Down feathers, 108, *108*
Dvinia, 122

E

Earth Science Integration, 41, 95
Earthworms, 43–45, *43, 44, 45*, 47, *act.* 62–63; behavior of, *act.* 149
Echinoderms, 58–61; characteristics of, 58, *58;* origin of, 61, *61;* types of, 59–60, *59, 60;* value of, 61
Ectotherm, 75, *act.* 76, 90
Eels, 79
Eggs, 106, *106*, 126–127, *126, 127;* amniotic, **91,** *91*
Elasmosaurus, 94, *94*
Elk, *116*
Endoskeleton, 73
Endotherm, 75, *act.* 76
Energy: animal adaptations for obtaining, 9, *9*
Environmental Science Integration, 109
Estivation, 85, 147
Excretory system: of earthworms, 44, *44;* of insects, 49
Exoskeleton, 48
Explore Activity, 7, 37, 71, 105, 133
Extinction, 94, *94*, 95; of mollusks, 41
Eye: compound, 49, *49*

F

Fan worms, 45, *45*
Farming. *see* Agriculture
Feathers, 108, *108*
Field Guides: Insects Field Guide, 162–165; Feline Traits Field Guide, 166–169
Fin, 73
Fish, 77–84; adaptations of, 9, *9*, 10, *11;* body systems of, 78, *78;* characteristics of, 77–79, *77, 78, 79;* feeding adaptations of, 79, *79;* fins of, 73; gas exchange in, 78, *78*, 81, *act.* 96–97; importance of, 84; origin of, 75, *75*, 84; scales of, 73, *73;* swim bladder of, 81; types of, 74, *74*, 75, 80–83, *80, 81, 83;* water temperature and respiration rate of, *act.* 96–97
Flagella, 15, *15*
Flatworms, 22–24, *23, 24, 26, act.* 28–29
Flight adaptations, 107–109, *107, 108, 109*
Flounder, 10, *11*
Flukes, 23, *23*
Foldables, 7, 37, 71, 105, 133
Fossil(s): of dinosaurs, 75
Freeliving organism, 22, *23*
Freshwater ecosystems: wetlands, *70*
Frogs, 86, *86*, 87, *87*, 88, *88*, 89

G

Gas exchange: in fish, 78, *78*, 81, *act*: 96–97
Gastropods, 39, *39*
Gestation period, 119
Gill(s), 38, 78, *78*
Gill slits, 73, *73*

Index

Gizzard(s): *act.* 105; of earthworms, 44, *44*

H

Habitats: *act.* 150–151, *150, 151*
Hagfish, 80, *80*
Halibut, 78, *78*
Handbooks. *see* Math Skill Handbook, Reference Handbook, Science Skill Handbook, and Technology Skill Handbook
Hawks, 9, *9*
Health Integration, 9, 51, 80, 135
Herbivores, 9, *9,* 115, *115,* 116
Hermaphrodites, 16, 45
Hibernation, 85, **147,** *147*
Horse, 115, *115*
Hummingbird, 111, *111,* 126

I

Ichthyosaur, 94, *94*
Imprinting, 137, *137*
Incubate, 106
Innate behavior, 134–135, *134, 135*
Insect(s), 49–51; abdomen of, 49; communication among, 142, *142;* controlling, 56; diseases and, 51; head of, 49, *49;* mandibles of, 51, *51;* metamorphosis of, 50, *50;* migration of, 148, *148;* spiracles of, 49; success of, 51; thorax of, 49
Insecticides, 56
Insight, 139, *139*
Instinct, 136, *136,* 140
Internet. *see* Science Online and Use the Internet
Invertebrate animals, 12; cnidarians, 17–20, *17, 18, 19, 20, act.* 21; segmented worms, 43–47, *43, 44, 45, 46, act.* 62–63; sponges, 14–16, *14, 15, 16*

J

Jawless fish, 74, *74,* 80
Jellyfish, 17, *17*

K

Kangaroo, 105, 118

L

Labs. *see* Activities, MiniLABs, and Try at Home MiniLABs
Lampreys, 80
Lancelets, 84, *84*
Larva, 16, *16,* 19
Law(s): Newton's third law of motion, 41
Learned behavior, 136–139, *136, 137, 138, 139*
Leeches, 46, *46,* 47
Lichens, 10
Light communication, 144–146, *145*
Lizards, 92, *92*
Lobe-finned fish, 75, *77,* 82
Lobsters, *54*
Lungfish, 83, *83*
Lyme disease, 27

M

Mammals, 74, *74,* 104, *104,* **114**–122; body systems of, 116–117, *116;* characteristics of, 114–116, *114, 115;* glands of, 114; hair of, 116; importance of, 122; origin of, 122, *122;* reproduction of, *104, 114,* 117, *117,* 119, *119;* skin of, 114; teeth of, 115, *115;* types of, 118–121, *118, 119, 120, 121*
Mammary glands, 114
Manatee, 119, *119*
Mandibles, 51, *51*
Mantle: of mollusk, 38
Marine worms, 45, *45,* 47
Marsupials, 118, *118*
Math Skill Handbook, 188–194
Math Skills Activities, 25, 2, 82
Medicine: leeches in, 46, *46,* 47
Medusa, 17, *17,* 19, *19*
Metamorphosis, 50, *50,* 86–87, *86–87*
Microbes, 9
Migration, 148, *148*
Millipedes, 53, *53*
Mimicry, 10, *10*
MiniLABs: Observing Planarian Movements, 24; Observing Metamorphosis, 50; Describing Frog Adaptations, 88; Modeling Feather Preening, 108; Observing Conditioning, 138
Mites, 53
Model and Invent, 150–151
Mollusks, 38–42; bivalves, 40, *40;* cephalopods, 40–41, *40, 41;* characteristics of, 38, *38;* classifying, 39–41; extinction of, 41; gastropods, 39, *39;* origin of, 41; segmented worms and, 47; shells of, *act.* 37, 42, *42;* univalves, 39, *39;* value of, 42, *42*
Molting, 48
Monotremes, 118, *118*
Moose, *114*
Mosasaur, 94, *94*
Mosquitoes, *51*
Motion: Newton's third law of, 41

N

National Geographic Visualizing: Parasitic Worms, 26; Arthropod Diversity, 54; Extinct Reptiles, 94; Birds, 112; Bioluminescence, 145
Nature of Science: Monarch Migration, 2–5
Nematodes, 27, *27*
Nerve cord, 73, *73*
Nerve net, 18
Nervous system: of earthworms, 45; of mammals, 116
Nest, 106
Nest building, 135, *135*
Newton's third law of motion, 41

Index

Notochord, 72, *73*

O

Octopus, 11, *11*, 40, 42
Omnivores, 9, *9*, **115**, *115*
Oops! Accidents in Science, 152–153
Open circulatory system, 38
Opossum, *118*
Ostrich, *126*
Owl, 111, *111*, 146, *146*
Oysters, 40, 42, *42*

P

Pack behavior, 134, *134*, 141
Parasites, 23, *23*, 24, *24*, *act.* 28–29
Parrotfish, *79*
Pearls, 42, *42*
Pharynx, 22
Pheromone, 143
Physical adaptations, 10, *10*
Pillbugs, 10, *10*, 55, *55*
Placenta, 119, *119*
Placentals, 119–121, *119*, *120*, *121*
Placoderms, 75, 84
Planarians, 22–23, *23*, 24
Plantlike animals, 8
Plesiosaur, 94, *94*
Polychaetes, 45, *45*
Polyp, 17, *17*, 19
Porpoise, *117*, 121
Postanal tail, 72, *73*
Predator adaptations, 11, *11*
Preening, 108
Primates, 120, *120*
Problem-Solving Activities, 117, 147
Protoavis, 113, *113*

R

Rabbits, 121, *121*
Radial symmetry, **13**, *13*
Radula, 39
Ray-finned fish, 83, *83*
Reef, 20, *20*
Reference Handbook, 195–201
Reflex, 135
Regeneration, 16, 55
Reproduction: of amphibians, 86; asexual, 16, 19, *19*, 23, *23*; of birds, 106, *106*; of earthworms, 45; of fish, 79; of mammals, 104, 114, 117, *117*, 119, *119*; of reptiles, 91, *91*, 93; sexual, 16, *16*, 19, *19*
Reptiles, 70, 74, **74**, 90–95; alligators, 93; characteristics of, 90–91, **90, 91**; crocodiles, 90, 93, **93**; extinct, 94, **94**, 95; importance of, 95; lizards, 92, *92*; origin of, 94, **94**, 95; snakes, *act.* 71, 91, 92, **92**; turtles, 91, 93, **93**, 95; types of, 91–94, **92, 93, 94**
Respiration: of amphibians, 85; of earthworms, 44; of fish, 78, *78*, *act.* 96–97
Respiratory system: of birds, 110, *110*; of mammals, 116
Roundworms, 25–27, *25*, *26*, *27*

S

Salamanders, 85, *85*, 86, 88
Sand dollars, 60, *60*
Sawfish, *79*
Scales, 73, *73*
Scallops, 40, *40*
Scavengers, 9
Science and History. *see* TIME
Science and Language Arts, 64–65,
Science and Society. *see* TIME
Science Online: Collect Data, 18; Research, 12, 46, 60, 87, 91, 110, 119, 137, 146
Science Skill Handbook, 170–183
Science Stats, 126–127
Schistosomiasis, 23
Scorpions, 52
Sea anemone, 17
Sea cucumbers, 60, *60*
Sea horses, 79, 83, *83*
Sea stars, 58, *58*, 59
Sea urchins, 60, *60*
Segmented worms, 43–47; characteristics of, 43; earthworms, 43–45, *43*, *44*, *45*, 47, *act.* 62–63; food eaten by, *act.* 62–63; leeches, 46, *46*, 47; marine worms, 45, *45*, 47; mollusks and, 47; origin of, 47, *47*; value of, 47
Sessile, 15
Setae, 43, 45
Sexual reproduction, 16, *16*, 19, *19*
Sharks, 73, 79, *79*, 80
Sheep, *132*
Shells: of mollusks, *act.* 37, 42, *42*
Shipworms, 42
Skin, 114
Skinks, *90*
Slugs, 39, *39*, 42
Snails, 39, *39*, 42
Snakes, 10, *10*, *act.* 71, 91, 92, *92*
Social behavior, 140–141, *140*, *141*
Society, 141
Soil: and earthworms, 47, 62
Sound communication, 144, *144*
Species: calculating number of, 25
Spicules, 15, *15*, 16
Spiders, 53, *53*, 54, 136, *136*
Spiracles, 49
Sponges, 14–16, *14*, *15*, *16*
Spongin, 15
Squid, 40, 41, *41*
Standardized Test Practice, 35, 69, 103, 131, 157, 158–159
Stinging cells, 18, *18*
Submission, 142, *142*
Swim bladder, 81
Swimmerets, 55
Symmetry: *act.* 7, 12–13; bilateral, **13,** *13*; radial, **13,** *13*

T

Tadpoles, 86, *86*
Tapeworms, 24, *24*
Technology Skill Handbook, 184–187

Index

Teeth: of mammals, 115, *115*
Tentacles, 18, *18*
Termites, 141, *141*
Territory, 141, *141*
Test Practices. *see* Standardized Test Practice
The Princeton Review. *see* Standardized Test Practice
Ticks, 53
Tiger, 11, 115, *115*, 141, *141*
TIME: Science and History, 30–31; Science and Society, 98–99
Traditional Activity, 62–63
Tree frogs, 86, *86*
Trial and error, 137, *137*
Trilobite, *56*
Trout, 10, *11*
Try at Home MiniLABs: Modeling Animal Camouflage, 10; Modeling the Strength of Tube Feet, 59; Modeling How Fish Adjust to Different Depths, 80; Inferring How Blubber Insulates, 116; Demonstrating Chemical Communication, 143
Tube feet, 58, *58*, 59
Turtle(s), 10, 91, 93, *93*, 95

Umbilical cord, 119
Univalves, 39, *39*
Use the Internet, 124–125

Venn diagram, 105
Vertebra, 73, *73*
Vertebrate animals, 12; characteristics of, 73, *73*; origins of, 75, *75*; types of, 74, *74, 75, act.* 76

Visualizing. *see* National Geographic

Water-vascular system, 58, *58*
Wetlands, 70
Whale, 11
Wings, 109, *109*
Wolves, 11, 141, *142*
Worms, 22–29, *22*. *see* Segmented worms; flatworms, 22–24, *23, 24, 26, act.* 28–29; free-living, **22,** *23, act.* 28–29; parasitic, 23, *23,* 24, *24, act.* 28–29; roundworms, 25–27, *25, 26, 27*

Zebras, 140, *140*

Credits

Art Credits

Glencoe would like to acknowledge the artists and agencies who participated in illustrating this program: Absolute Science Illustration; Andrew Evansen; Argosy; Articulate Graphics; Craig Attebery represented by Frank & Jeff Lavaty; CHK America; Gagliano Graphics; Pedro Julio Gonzalez represented by Melissa Turk & The Artist Network; Robert Hynes represented by Mendola Ltd.; Morgan Cain & Associates; JTH Illustration; Laurie O'Keefe; Matthew Pippin represented by Beranbaum Artist's Representative; Precision Graphics; Publisher's Art; Rolin Graphics, Inc.; Wendy Smith represented by Melissa Turk & The Artist Network; Kevin Torline represented by Berendsen and Associates, Inc.; WILDlife ART; Phil Wilson represented by Cliff Knecht Artist Representative; Zoo Botanica.

Photo Credits

Abbreviation Key: Abbreviation key: AA=Animals Animals; AH=Aaron Haupt; AMP=Amanita Pictures; BC=Bruce Coleman, Inc.; CB=CORBIS; DM=Doug Martin; DRK=DRK Photo; ES=Earth Scenes; FP=Fundamental Photographs; GH=Grant Heilman Photography; IC=Icon Images; KS=KS Studios; LA=Liaison Agency; MB=Mark Burnett; MM=Matt Meadows; PE=PhotoEdit; PD=PhotoDisc; PQ=PictureQuest; PR=Photo Researchers; SB=Stock Boston; TSA=Tom Stack & Associates; TSM=The Stock Market; VU=Visuals Unlimited.

Cover Bill Bachmann/Rainbow; **v** John Cancalosi/DRK; **vi** Michael Newman/PE; **vii** Bill Kamin/VU; **2** (t)Fred Habegger from GH, (b)Fritz Polking/VU; **3** (t)DM, (b)Richard Megna/FP; **4** David Woodfall/DRK; **5** (t)Michael Newman/PE, (b)James L. Amos/CB; **6** Bruna Stude/Omni-Photo Communications; **6-7** A. Witte/C. Mahaney/Stone; **7** IC; **8** Zig Leszczynski/AA; **9** (l)Jeff Foott/DRK, (c)Leonard Lee Rue III/DRK, (r)Hal Beral/VU; **10** (t)Ken Lucas/VU, (bl)Joe McDonald/VU, (br)Zig Leszczynski/AA; **11** (tl)Tom J. Ulrich/VU, (tc)Peter & Beverly Pickford/DRK, (tr)Michael Fogden/DRK, (b)Stuart Westmoreland/Mo Yung Productions; **12** (l)Stephen J. Krasemann/DRK, (r)Ford Kristo/DRK; **14** (l)Andrew J. Martinez/PR, (r)Glenn Oliver/VU; **17** (t)Norbert Wu/DRK, (bl)Fred Bavendam/Minden Pictures, (br)H. Hall/OSF/AA; **18** Gerry Ellis/ENP Images; **20** David B. Fleetham/VU; **21** Larry Stepanowicz/VU; **23** (tl)T.E. Adams/VU, (tr)Science VU/VU, (b)Oliver Meckes/Eye of Science/PR; **24** Triarch/VU; **25** Oliver Meckes/Ottawa/PR; **26** (t)NIBSC/Science Photo Library/PR, (cl)Sinclair Stammers/Science Photo Library/PR, (cr)Arthur M. Siegelman/VU, (b, l to r)Oliver Meckes/PR, Andrew Syred/Science Photo Library/PR, Eric V. Grave/PR, Cabisco/VU; **27** R. Calentine/VU; **28** (t)T.E. Adams/VU, (b)Daemmrich Photography; **29** MM; **30 30-31** PD; **31** (l)Thayer Syme/FPG, (r)Shirley Vanderbilt/Index Stock; **32** (t)Kjell B. Sandved/VU, (bl br)R. Calentine/VU; **33** (l)Runk/Schoenberger from GH, (r)James H. Robinson/AA; **35** Donald Specker/AA; **36** Meckes/Ottawa/Eye of Science/PR; **36-37** Frans Lanting/Minden Pictures; **37** John Evans; **38** Wayne Lynch/DRK; **39** (l)Jeff Rotman Photography, (r)James H. Robinson/AA; **40** Joyce & Frank Burek/AA; **41** Clay Wiseman/AA; **42** Bates Littlehales/AA; **43** Beverly Van Pragh/Victoria Museum; **44** Donald Specker/AA; **45** (t)Charles Fisher, Penn State University, (bl)Mary Beth Angelo/PR, (br)Kjell B. Sandved/VU; **46** St. Bartholomew's Hospital/Science Photo Library/PR; **48** Tom McHugh/PR; **49** (t)Ted Clutter/PR, (b)Kjell B. Sandved/VU; **52** Lynn Stone; **53** (l)Bill Beatty, (r)Patti Murray/AA; **54** (tl)Bill Beatty/Wild & Natural, (tc)Robert F. Sisson, (tr)Lynn Stone, (c)Brian Gordon Green, (c)Joseph H. Bailey/National Geographic Image Collection, (cr)Jeffrey L. Rotman/CB, (b)Timothy G. Laman/National Geographic Image Collection; **55** (t)James P. Rowan/DRK, (b)Leonard Lee Rue/PR; **56** Ken Lucas/VU; **57** TSA; **58** Scott Smith/AA; **59** Clay Wiseman/AA; **60** (tl)Andrew J. Martinez/PR, (tr)David Wrobel/VU, (b)Gerald & Buff Corsi/VU; **61** Ken Lucas/VU; **62 63** MM; **64** (t)David M. Dennis, (b)Harry Rogers/PR; **65** Mark Lennihan/AP; **66** (tl)Hal Beral/VU, (tr)Leroy Simon/VU, (bl)PR, (br)Joyce & Frank Burek/AA; **67** (l)Charles McRae/VU, (r)Mark Moffet/Minden Pictures; **68** William Leonard/DRK; **70** Jean Hall/Holt Studios/PR; **70-71** Robert Lubeck/AA; **71** Laura Sifferlin; **72** Fred Bavendam/Minden Pictures; **73** Omni-Photo Communications; **74** (t to b)H. W. Robison/VU, Nicklin/Minden Pictures, Flip Nicklin/Minden Pictures, John M. Burnley/PR, George Grall/National Geographic, M. P. Kahl/DRK, Grace Davies/Omni-Photo Communications; **75** T. A. Wiewandt/DRK;

76 IC; **77** (l)Meckes/Ottawa/PR, (cl)Rick Gillis/University of Wisconsin-La Crosse, (cr r)Runk/Schoenberger from GH; **78** Ken Lucas/VU; **79** (tl)James Watt/AA, (tc)Norbert Wu/DRK, (tr)Fred Bavendam/Minden Pictures, (b)Richard T. Nowitz/PR; **80 82** Tom McHugh/PR; **83** (t)Tom McHugh/Steinhart Aquarium/PR, (bl)Bill Kamin/VU, (bc)Norbert Wu/DRK, (br)Michael Durham/Gerry Ellis/ENP Images; **84** Runk/Schoenberger from GH; **85** Fred Habegger from GH; **86** (t)David Northcott/DRK, (bl br)Runk/Schoenberger from GH; **87** (l)Runk Schoenberger from GH, (r)George H. Harrison from GH; **88** (l)Mark Moffett/Minden Pictures, (r)Michael Fogden/DRK; **89** Lynn M. Stone; **90** Joe McDonald/VU; **92** (l)Klaus Uhlenhut/AA, (r)Rob & Ann Simpson/VU; **93** (t)Mitsuaki Iwago/Minden Pictures, (c)Belinda Wright/DRK, (b)G. & C. Merker/VU; **94** (tl)John Sibbick, (tr)Karen Carr, (c)Chris Butler/SPL/PR, (bl)Jerome Connolly, courtesy The Science Museum of Minnesota, (br)Chris Butler; **96** (t)Steve Maslowski/VU, (b)Michael Newman/PE; **97** KS; **98** (t)David Aubrey/TSM, (b)Carl Roessler/FPG; **99** (tl)Michael Fogden/AA, (tr)Tim Flach/Stone, (b)R. Rotolo/LA; **100** (tl)Marty Cordano/DRK, (tr)Tim Fitzharris/Minden Pictures, (bl)Runk/Schoenberger from GH, (br)John Cancalosi/DRK; **101** (l)Des & Jen Bartlett/National Geographic, (r)David Northcott/DRK; **104** Des & Jen Bartlett/National Geographic; **104-105** Stephen J. Krasemann/DRK; **105** IC; **106** Michael Habicht/AA; **108** (l)Crown Studios, (r)KS; **109** (l)Lynn Stone/AA, (r)Arthur R. Hill/VU; **111** (l)Zefa Germany/TSM, (r)Sid & Shirley Rucker/DRK; **112** (tl)Wayne Lankinen/DRK, (tr)Ron Spomer/VU, (c)Kennan Ward/CB, (bl)Steve Maslowski, (br)M. Philip Kahl/Gallo Images/CB, (br)Rod Planck/PR; **114** Stephen J. Krasemann/DRK; **115** (t)Gerard Lacz/AA, (bl)Tom Brakefield/DRK, (br)John David Fleck/LA; **116** Bob Gurr/DRK; **117** Amos Nachoum/TSM; **118** (t)Jean-Paul Ferrero/AUSCAPE, (bl)Phyllis Greenberg/AA, (br)John Cancalosi/DRK; **119** (t)Carolina Biological Supply/PhotoTake NYC, (b)Doug Perine/DRK; **120** (t to b)Stephen J. Krasemann/DRK, David Northcott/DRK, Zig Leszczynski/AA, Ralph Reinhold/AA, Anup Shah/AA, Mickey Gibson/AA; **121** (t to b)Fred Felleman/Stone, Robert Maier/AA, Tom Bledsoe/DRK, Wayne Lynch/DRK, Joe McDonald/AA, Kim Heacox/DRK; **124** (t)Wayne Lankinen/DRK, (c)David Welling/AA, (b)Richard Day/AA; **125** MM; **126** (t)MB, (cl)Bob & Clara Calhoun/BC/PQ, (cr)Joe McDonald/AA, (b)Jeff Fott/DRK; **127** (l)Christie's Images, London/Bridgeman Art Library, (c)John Welzenbach/TSM, (b)MB; **128** (tl)Howie Garber/AA, (tr)Bob Gurr/DRK, (bl)Johnny Johnson/DRK, (br)Hans & Judy Beste/AA; **129** (tl tr)Tom & Pat Leeson/DRK, (b)Tom Brakefield/DRK; **132** Gary W. Carter/VU; **132-133** Robert Mackinlay/Peter Arnold, Inc.; **133** MB; **134** (l)Michel Denis-Huot/Jacana/PR, (r)Zig Lesczynski/AA; **135** (l)Jack Ballard/VU, (c)Anthony Mercieca/PR, (r)Joe McDonald/VU; **136** (t)Stephen J. Krasemann/Peter Arnold, Inc., (b)Leonard Lee Rue/PR; **137** (t)The Zoological Society of San Diego, (b)Margret Miller/Photo Reseachers; **140** Michael Fairchild; **141** (t)Bill Bachman/PR, (b)Fateh Singh Rathore/Peter Arnold, Inc.; **142** Jim Brandenburg/Minden Pictures; **143** Michael Dick/AA; **144** (l)Richard Thorn/VU, (c)Arthur Morris/VU, (r)Jacana/PR; **145** (tl)T. Frank/Harbor Branch Oceanographic Institution, (bl bc)Peter J. Herring, (others)Edith Widder/Harbor Branch Oceanographic Institution; **146** Stephen Dalton/AA; **147** Richard Packwood/AA; **148** Ken Lucas/VU; **150** (t)Dave B. Fleetham/TSA, (b)Gary Carter/VU; **151** The Zoological Society of San Diego; **152** Walter Smith/CB; **152-153** Bios (Klein/Hubert)/Peter Arnold, Inc.; **153** Courtesy The Seeing Eye; **154** (t)Norbert Wu/Peter Arnold, Inc., (tr)Fritz Prenzel/AA, (b)Remy Amann-Bios/Peter Arnold, Inc.; **155** (l)Valerie Giles/PR, (r)J & B Photographers/AA; **156** Alan & Sandy Carey/PR; **160-161** PD; **162** (t)PR, (b)David M. Dennis; **163** (t)Roy Morsch/TSM, (cl)Harry Rogers/PR, (cr)Donald Specker/AA, (b)Roger K. Burnard; **164** (t)Tom McHugh/PR, (tr)Donald Specker/AA, (cl)Harry Rogers/PR, (c)Carroll W. Perkins/AA, (cr)Patti Murray/AA, (b)Donald Specker/AA; **165** (tl)Harry Rogers/PR, (tr)Ken Brate/PR, (cl)James H. Robinson/PR, (cr)Linda Bailey/AA, (bl)Ed Reschke/Peter Arnold, Inc., (br)MM; **166** (t)Carolyn A. McKeone/PR, (b)J. & P. Wegner/AA; **167** (tl)Yann Arthus-Bertrand/CB, (tr)Barbara Reed/AA; **167** (bl)Chanan Photography, (br)J-L Klein & M-L Hubert/OKAPIA/Photo Resarchers; **168** (tl)Renee Stockdale/AA, (tr)Joan Baron/PR, (bl)Carolyn A. McKeone/PR, (br)Yann Arthus-Bertrand/CB; **169** (tl)Stephen Green/TSM, (tr)J. & P. Wagner/AA, (bl)Jane Howard/PR, (br)Ulrike Schanz/AA; **170** Timothy Fuller; **174** First Image; **177** Dominic Oldershaw; **178** StudiOhio; **179** First Image; **181** Richard Day/AA; **184** Paul Barton/TSM; **187** Charles Gupton/TSM; **197** MM; **198** (t)NIBSC/Science Photo Library/PR, (bl)Dr. Richard Kessel, (br)David John/VU; **199** (t)Runk/Schoenberger from GH, (bl)Andrew Syred/Science Photo Library/PR, (br)Rich Brommer; **200** (t)G.R. Roberts, (bl)Ralph Reinhold/ES, (br)Scott Johnson/AA; **201** Martin Harvey/DRK.

CREDITS C ◆ 217

PERIODIC TABLE OF THE ELEMENTS

Columns of elements are called groups. Elements in the same group have similar chemical properties.

Element — Hydrogen
Atomic number — 1
Symbol — H
Atomic mass — 1.008

State of matter

Each element has a block in the periodic table. Within a block, you can find important information about the element.

Group	1	2	3	4	5	6	7	8	9
1	Hydrogen 1 H 1.008								
2	Lithium 3 Li 6.941	Beryllium 4 Be 9.012							
3	Sodium 11 Na 22.990	Magnesium 12 Mg 24.305							
4	Potassium 19 K 39.098	Calcium 20 Ca 40.078	Scandium 21 Sc 44.956	Titanium 22 Ti 47.88	Vanadium 23 V 50.942	Chromium 24 Cr 51.996	Manganese 25 Mn 54.938	Iron 26 Fe 55.847	Cobalt 27 Co 58.933
5	Rubidium 37 Rb 85.468	Strontium 38 Sr 87.62	Yttrium 39 Y 88.906	Zirconium 40 Zr 91.224	Niobium 41 Nb 92.906	Molybdenum 42 Mo 95.94	Technetium 43 Tc 97.907	Ruthenium 44 Ru 101.07	Rhodium 45 Rh 102.906
6	Cesium 55 Cs 132.905	Barium 56 Ba 137.327	Lanthanum 57 La 138.906	Hafnium 72 Hf 178.49	Tantalum 73 Ta 180.948	Tungsten 74 W 183.84	Rhenium 75 Re 186.207	Osmium 76 Os 190.2	Iridium 77 Ir 192.22
7	Francium 87 Fr 223.020	Radium 88 Ra 226.025	Actinium 89 Ac 227.028	Rutherfordium 104 Rf (261)	Dubnium 105 Db (262)	Seaborgium 106 Sg (263)	Bohrium 107 Bh (262)	Hassium 108 Hs (265)	Meitnerium 109 Mt (266)

Rows of elements are called periods. Atomic number increases across a period.

The arrow shows where these elements would fit into the periodic table. They are moved to the bottom of the page to save space.

Lanthanide series

Cerium 58 Ce 140.115	Praseodymium 59 Pr 140.908	Neodymium 60 Nd 144.24	Promethium 61 Pm 144.913	Samarium 62 Sm 150.36

Actinide series

Thorium 90 Th 232.038	Protactinium 91 Pa 231.036	Uranium 92 U 238.029	Neptunium 93 Np 237.048	Plutonium 94 Pu 244.064